|        |  DATE  |        |        |
| ------ | ------ | ------ | ------ |
|        |        |        |        |
|        |        |        |        |
|        |        |        |        |
|        |        |        |        |
|        |        |        |        |
|        |        |        |        |
|        |        |        |        |
|        |        |        |        |
|        |        |        |        |
|        |        |        |        |
|        |        |        |        |
|        |        |        |        |
|        |        |        |        |

# Neurophysics

# NEUROPHYSICS

**Alwyn C. Scott**

Department of Electrical and Computer Engineering
The University of Wisconsin
Madison, Wisconsin

**A Wiley-Interscience Publication**
**JOHN WILEY & SONS**
New York · London · Sydney · Toronto

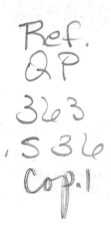

*Library of Congress Cataloging in Publication Data:*

Scott, Alwyn, 1931-
Neurophysics.

"A Wiley-Interscience publication."
  Bibliography: p.
  Includes index.
  1.  Neurophysiology.     2.   Biological physics.
3.   Action potentials (Electrophysiology)   I.   Title.

QP363.S36           612'.8'043           77-2762
ISBN 0-471-02998-X

Printed in the United States of America

10  9  8  7  6  5  4  3  2  1

# Preface

Although physical scientists have contributed actively to early neural studies, current neuroscience is dominated by investigators trained in the life sciences such as anatomy, physiology, biochemistry, zoology, psychology, and neurology. It is, of course, entirely appropriate that physical scientists play only a minor role in the study of the living brain, but we should not retire altogether from the stage. We still have contributions of value to make. The main purpose of the present work is to sketch the vast fields of neuroscience from the perspective of a physical scientist and to indicate suitable problem areas for physicists, engineers, and applied mathematicians. But it is hoped and expected that the book may also be of interest to life scientists who seek key ideas for understanding the mechanisms of nature. An extensive bibliography is included to stimulate this interest.

There are dangers facing the physical scientist who follows current trends toward an interest in the life sciences. The first of these is a lack of humility in the face of the enormous dynamic complexity exhibited by living systems. We physical scientists tend to assume that our reductive and analytical tools are universally applicable. The present author rejects this assumption for reasons that I attempt to clarify in Chapter 8, and suggests instead that we revere the immense variety displayed by physiological material. Furthermore, I must protest my ignorance of physiology and offer apologies to those to whom this is obvious. But I suggest, on the other hand, that there are areas of neuroscience in which the standard notions of physical science and applied mathematics can help. Foremost among these are studies of nonlinear diffusion on single nerve fibers that lead to an appreciation of the nerve impulse as an independent dynamic entity. Extensions of these studies indicate that the dynamic behavior of an individual neuron may be considerably more complex than has previously been assumed. Chapters 4 through 6 carry

v

the discussion from single-fiber propagation to a sketch of the "multiplex neuron", and form the backbone of the book.

Applied mathematicians work in diverse areas of science and are in a position to appreciate the hierarchical organization of scientific activity. This book is organized according to hierarchical levels not simply for convenient exposition but also to emphasize the importance of such levels as a key feature of scientific knowledge. The nerve impulse is not only an important element in understanding neuron dynamics. Arising as an "elementary particle of thought" out of an underlying partial differential equation for nonlinear diffusion, it serves as a paradigm for the way that a relatively simple and continuous description at one level can provide atomistic entities that serve as a basis for describing dynamic activity at higher levels. This notion is emphasized in Chapter 7. In attempting to bridge the long, unexplored chasm of dynamic possibilities separating the facts of neurology from the facts of psychology, Hebb has suggested the "cell assembly" as an intermediate atomistic entity that can provide a basis for a description of thought. We should not waste time haggling over the "reality" of such entities in terms of physics and chemistry. They can be as "real" as a tornado (arising out of the hydrodynamic equation describing the atmosphere) or an electron (arising out of some as yet undiscovered nonlinear version of electromagnetic theory). As one begins to appreciate Hebb's ideas, much of the philosophical distinction between "realism" and "idealism" becomes irrelevant.

At a much more basic level, the dynamic behavior of an active nerve membrane is yet to be satisfactorily explained in terms of the underlying biochemistry and electrochemistry. Both the theoretical and experimental tools of physics seem likely to contribute to the solution of this outstanding riddle. Chapter 3 surveys the current state of this work and provides a basis for the nerve fiber studies in Chapter 4, whereas Chapter 2 discusses the corresponding electromagnetic problem.

This book has evolved out of notes for a course given each year since 1971 in the Department of Electrical and Computer Engineering at the University of Wisconsin. Students in the course have appeared from the Departments of Computer Science, Mathematics, Physiology, Physics, and Zoology, in addition to Engineering. One important (and somewhat unexpected) pedagogical value of such a course has been the horizontal flow of information, insights, and attitudes between all participants. I hope that a similar interaction may be encouraged on a wider scale by publication of these notes.

Finally I am happy to have the opportunity to express appreciation for all the assistance that has made publication possible. Generous support from the National Science Foundation over the past 15 years (for nonlinear wave studies) has been of fundamental importance, as has a year of full-time effort during 1974–1975 supported jointly by the National Library of Medicine and the Mathematics Research Center of the University of Wisconsin. During 1969–1970 the present author was privileged to spend a research year at the *Laboratorio di Cibernetica*, in Naples, Italy which, under the leadership of Professor E. R. Caianiello, is devoted to the interdisciplinary problems of neuroscience. The ideas and attitudes gathered during this year established a basis for the entire project, and many friends at the *Laboratorio* have continued to provide both help and encouragement. During the autumn semester (1976), I have had the opportunity to present a course based on these notes in the Mathematics Department at the University of Arizona. The interested criticism offered by students and staff has been invaluable in the final editing of the manuscript.

ALWYN C. SCOTT

*Madison, Wisconsin*
*April 1977*

# Acknowledgments

The preparation of this book has been aided by many people. Several figures and quotations have been made available through the courtesy of their original authors and publishers. Chapter 2 appeared in the *Reviews of Modern Physics* (April 1975); Chapters 7 and 8 are a revised and expanded version of a study that originally appeared as a Mathematics Research Center Report (No. 1548, October 1975) and was later published in the *Journal of Mathematical Psychology* (May 1977). Specific acknowledgment is made for the following items.

Figs. 1-3 and 3-13.   Dr. Kenneth S. Cole.
Fig. 1-5.   Dr. William Feindel for the Montreal Neurological Hospital and Institute.
Fig. 3-1.   Professor David E. Green.
Fig. 3-5.   Professor Carver Mead.
Figs. 3-8, 4-2, and 4-3.   Professor Andrew Huxley and the *Journal of Physiology*.
Figs. 3-9 and 3-12.   Dr. Kenneth S. Cole and the Regents of the University of California.
Fig. 3-14.   Dr. David Landowne and the *Journal of Physiology*.
Fig. 4-4.   Drs. James W. Cooley and F. A. Dodge, Jr., and the Rockefeller University Press.
Fig. 4-9.   Dr. Richard FitzHugh and the McGraw-Hill Book Company.
Fig. 4-10.   Dr. Richard FitzHugh and the American Physiological Society.
Figs. 4-14 and 5-1.   Dr. John Rinzel and the Rockefeller University Press.
Fig. 4-18.   Professor J. Z. Young and the Oxford University Press.
Fig. 4-22.   Dr. David Bod an.
Fig. 4-24.   Dr. Stephen Goldstein, Dr. W. Rall, and the Rockefeller University Press.
Poem, page 178.   The Viking Press and Faber & Faber Ltd.
Fig. 6-2a.   Dr. John N. Barrett, Dr. W. E. Crill, and the *Journal of Physiology*.
Fig. 6-2b.   Dr. Raphael L. Poritsky.
Fig. 6-3.   Dr. Stephen Waxman and the Elsevier/North Holland Biomedical Press.
Fig. 6-6.   Professor B. Katz and the *Journal of Physiology*.
Fig. 6-10.   Dr. E. J. Furshpan and the *Journal of Physiology*.
Figs. 6-20 and 6-21.   Dr. L. Tauc and the *Journal of General Physiology*.
Quotation, pages 196–197.   Professor J. Y. Lettvin.

Poem, page 228.   The MIT Press.
Fig. 7-3.   Grossett & Dunlap, Inc.
Fig. 7-4.   The Oxford University Press.
Fig. 7-6.   Professor T. Ishihara.
Quotation, page 235.   Theodore Roszak.
Fig. 7-11.   Professor R. L. Beurle and the Royal Society of London.
Fig. 7-12.   Professor Jack D. Cowan.
Quotation, page 281.   Charles Scribner's Sons.

                                                              ACS

# Contents

# 1

# Introduction

Wer will was Lebendig's erkennen und beschreiben,
Sucht erst den Geist heraus zu treiben,
Dann hat er die Teile in seiner Hand,
Fehlt leider nur das geistige Band.

      Goethe

A striking event in the early development of electrophysiology was the publication by Luigi Galvani in 1791 of his "Commentary on the forces of electricity and their relation to muscular motion" (Green, 1953). The effects of artificial, atmospheric, and animal electricities on nerve-muscle preparations from the frog were carefully discussed, leading him to the following query and response:

> Either nerves are of an insulative nature, as some surmise, and cannot then perform the function of conductors; or they are conductive: and how then could it be that the electric fluid should be contained within them and not be permitted to escape and diffuse to neighboring parts, not without great detrement surely of muscular contractions?

In response he offered:

> But this inconvenience and difficulty will easily be met by him who imagines the nerves so constituted that they are hollow within, or composed of some material suitable for conveying electric fluid, but externally they are either oily or are fused with some other substance which prevents the

1

effusion and dissipation of the said electric fluid running
through them.

This hypothesis is substantially correct, and much of the subsequent history of electrophysiology (Brazier, 1959; Harmon and
Lewis, 1966) has been devoted to working out the details.  Since
the main objective of the present work is to present some of those
details as they are currently understood, let me begin with an attempt to place in context the contributions that have led to an understanding of the nonlinear dynamics associated with propagation
of a voltage pulse, or "action potential," along a nerve fiber.

In 1850 Helmholtz used a cleverly designed apparatus (see
Fig. 1-1) to show that the signal velocity on a frog's sciatic nerve
is not immeasurably large, as had been previously assumed, but
some 27 m/sec.  The basic idea is both simple and elegant.  Clo-
sure of switch (V) simultaneously breaks the primary (P) initiating
a nerve pulse (N) and starts a time measurement on the ballistic
galvanometer (G).  When the muscle (M) twitches, a mercury contact at k is broken and the galvanometer stops.  The difference
of times measured for inputs at terminals (3-4) and (5-6) divided
into the corresponding distance along the nerve yields a velocity.
Bernstein (1868) described the details of an even more impressive
experimental tour de force; he measured the shape (potential plot-
against time) of the action potential on a frog's nerve and showed
that the velocity was equal to the signal velocity measured by
Helmholtz.  It is a fascinating experience to read these early papers and appreciate the experimental results that were obtained a
half century before Gasser and Erlanger (1922) introduced the
cathode-ray oscilloscope into electrophysiological research.

The basic problem was to understand the physical process involved in the propagation of the action potential.  Weber (1873 a,
b) took an important step with his fundamental study of the flow
of electricity in cylinders; indeed, we begin our analytic consideration with this calculation in Chapter 2.  Hermann (1879 b) seems
to have the correct physical ideas in mind.  He notes the similar-
ity of nerve propagation to a line of burning powder but rejects a
purely chemical explanation since this would seem to require activity throughout the entire cell.  He describes circulating currents that excite the neighborhood of a pulse and indicates that
these equations would lead to a form of the "heat equation. " This
line of thought, he wrote in 1879, "genügt uberhaupt . . . der

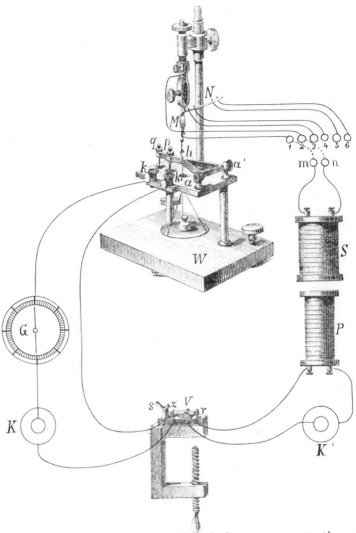

FIGURE 1-1.    Apparatus used by Helmholtz to measure the signal
velocity on a nerve fiber (Hermann, 1879a).

gestéllten Aufgabe nicht. " Hermann did not appreciate the descriptive power of a nonlinear diffusion equation until later and
even then he felt such problems would lead to "enormous mathematical difficulties" (Hermann, 1905). By this time Bernstein
(1902), building on studies of charge transport in ionic solutions

by Nernst (1888, 1889) and Planck (1890a, b) had stated in his "membrane hypothesis" that the action potential was the dis-charge of a (Nernst) diffusion potential caused by an increase in ionic permeability of the membrane.

The concept of a nerve cell or "neuron" as an independently functioning unit was established through the extensive anatomical studies of Ramón y Cajal (1908), and a survey of this research prepared shortly before his death in 1934 is now available in English (Ramón y Cajal, 1954). Most neurons display an input branching structure of "dendrites" called "dendritic trees" an en-larged cell body, and an output fiber or "axon" that eventually branches into an axonal tree. If appropriate firing conditions are established at the dendritic inputs, the cell body sends a pulse outward on the axon. An idea of the variety of neurons that fall within this basic pattern may be obtained through reference to Fig. 1-2, which is a reprint of the 1906 Nobel lecture of Ramón y Cajal and indicates some of the cerebellar (or motor control) cir-cuitry in the central nervous system of vertebrates. A variety of tree shapes are observed, each presumably adapted for the func-tion of a particular cell. The size of nerve cells also varies widely. For example, the sciatic nerve of a giraffe contains ax-ons of several meters in length, and the giant axon of the squid can be a millimeter in diameter. In the present work, the term "nerve fiber" implies both axons and dendrites, although most of the available experimental data are for large axons.

The following decades saw: (a) demonstration of the "all-or-nothing" nature of nerve-fiber response to stimulation (Lucas, 1909; Adrian, 1914), (b) confirmation of the existence of the cell membrane and measurement of its electrical capacitance (Fricke, 1923), (c) discovery of the squid giant axon (Young, 1936), (d) demonstration that the membrane conductance of a squid giant axon increases during the action potential (Cole and Curtis, 1938, 1939 ; see Fig. 1-3), and (e) the observation by Cole (1949) that membrane voltage (rather than current) is the more useful depend-ent variable for a phenomenological description. The activities of these years are described in detail in the recent book by Cole (1968); part history, part careful scientific discussion, this book should be studied by everyone who wishes to understand twentieth-century electrophysiology. Finally, the pieces of the puzzle were put together in the brilliant works of Hodgkin and Huxley (1952). They showed how measurements of the conductive parameters of a

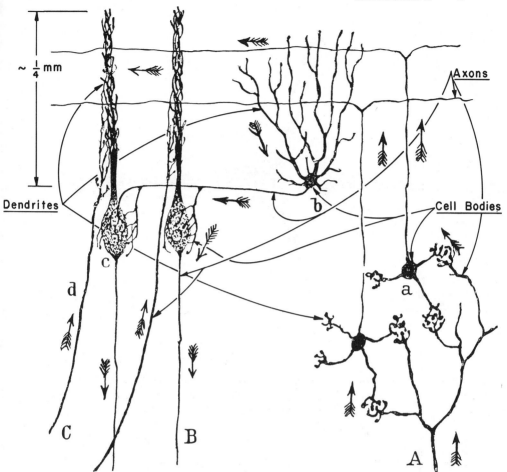

~ $\frac{1}{4}$ mm

Dendrites

Axons

Cell Bodies

FIGURE 1-2.   Semischematic diagram of neuron structure in the
cerebellum by Ramón y Cajal (1908): (A) mossy fiber
input; (a) granule cell; (B) Purkinje axon output;
(b) basket cell; (c) Purkinje cell body;
(C-d) climbing fiber input.

nerve fiber can be used to directly calculate both the shape and
the velocity of an action potential on the squid giant axon.

A great problem for modern thought, however, continues to be
that of understanding the human brain and, more generally, the
human personality in the context of natural philosophy.   Let's take

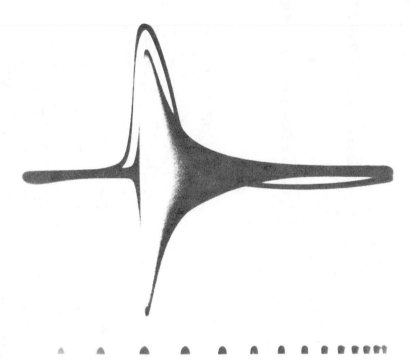

FIGURE 1-3.   Direct measurement of the increase in membrane
              conductance (band) during the action potential (line)
              on the squid giant axon;(Cole and Curtis, 1938).
              Time marks are 1 msec.

a look at it (Fig. 1-4).   A prominent feature is the reflection symmetry that conforms to the general bilateral symmetry of the body.
Each of two halves of the cerebral cortex contains neurons estimated at 5 billion to 9.2 billion in number, with a mean of 7.0
billion.   These observations were performed by six anatomists
since 1899 (Schade and Smith, 1970).   The study of such a complex living entity falls naturally into a hierarchy of scientific levels (Polanyi, 1962,1965; Koestler, 1971; Anderson, 1972) that
might be outlined as follows:

    1.   biochemistry from the chemical elements;
    2.   membrane electrodynamics from biochemistry;
    3.   neuron behavior from membrane dynamics and
         electromagnetic theory;

4.  neural network dynamics from neuron behavior;
5.  cell assemblies from network dynamics;
6.  psychology from cell assembly dynamics;
7.  human personality and culture from psychology.

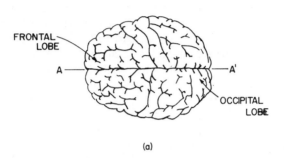

(a)

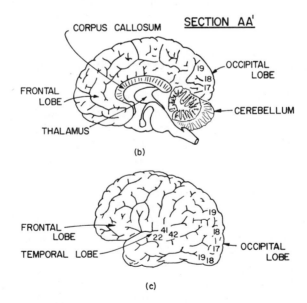

FIGURE 1-4.  The human central nervous system; (a) top view;
(b) section A-A'; (c) left-side view.

Level 3 is becoming recognized as a well-founded area for scientific study in the classical mode.   In the terminology of Kuhn (1962),the nonlinear diffusion equation developed by Hodgkin and Huxley is a widely accepted "paradigm" through which the complex dynamic properties of individual nerve cells are being systematically explored.   Such studies are the main subject of the present book.   They lead to the concept of the "multiplex neuron," which is more like a small computer than a single threshold logic element.

However, the mysteries of the cortex beckon us on to other levels of the scientific hierarchy.   An example of a higher-level experiment in electrophysiology is the inducement of experiential hallucinations by gentle electrical stimulation of the human cortex as summarized by Penfield and Perot (1963).   The purpose of such stimulation is to locate the focus of epileptic activity prior to sur-gical excision of a portion of the cortex.   In this study the records of 1288 consecutive brain operations for focal epilepsy were ex-amined.   Gentle electrical stimulation was applied to the temporal lobes of 520 patients, of whom 40 reported experiential responses. Stimulating currents between $50\mu A$ and $500\mu A$ were used in square waves of 2 to 5 msec at frequencies of 40 to 100 p. p. s.    A typi-cal example is:

Case 36. - M. M.   This 26-year-old woman had her first seizure at age 5.   Her attacks at first consisted of a sen-sation in one arm and leg followed by weakness in the leg, but when she was in college the pattern changed.   The pat-tern was:  (a)  interpretive illusion -- a feeling of familiar-ity and fear; (b)  experiential hallucination-- combined vis-ual and auditory; (c)  automatism.   She had sudden "flashes," which she described as experiencing something that she had experienced before.   She gave as examples: being under the grape arbour at her grandparents' farm (she felt she was there herself); sitting in the railroad station of a small town that was either Garrison or Vanceburg, Kentucky-- it was in the winter, the wind was blowing outside, and she was wait-ing for a train.

Dr. Feindel recorded one of her spontaneous ictal hallu-cinations as she recounted it to him at the close of an attack while in hospital:

She had the same flash-back several times.   These had to

do with her cousin's house or the trip there--a trip she had
not made for ten to fifteen years but used to make often as a
child.  She is in a motor car which had stopped before a rail-
way crossing.  The details are vivid.  She can see the swing-
ing light at the crossing.  The train is going by--it is pulled
by a locomotive passing from the left to right and she sees
coal smoke coming out of the engine and flowing back over
the train.  On her right there is a big chemical plant and she
remembers smelling the odor of the chemical plant.

The windows of the automobile seem to be down and she
seems to be sitting on the right side and in the back.  She
sees the chemical plant as a big building with a half-fence
next to the road.  There is a large flat parking space.  The
plant is a big rambling building-- no definite shape to it.
There are many windows.

Whether this is actually true or not she does not know
but it looks like that in the flashes.  She thinks she hears the
rumble of the train.  It is made up of black flat-top cars full
of cinders-- the kind they use for work on the road.  In another flash-back she says she sees her cousin's home and she is
in it.  She smells coffee-- "They always have coffee," and
"It is part of the atmosphere of the house. "

At operation the right temporal region was explored and
incusural sclerosis was found.  Figure 1-5a is a photograph
of the patient prepared for operation.  Local anesthetic has
been injected and the incision is outlined.  In Fig. 1-5b the
scalp, bone, and dura have been turned down and numbered
tickets mark the sites of positive cortical stimulations.  The
white thread indicates the area of temporal lobe to be removed.

The following experiential responses were produced with
stimulation (numbers correspond to points of electrical stimulation indicated in Fig. 1-5b):

11.   She said, "I heard something familiar, I do not know
what it was. "
11.   Repeated without warning.  "Yes, sir, I think I heard a
mother calling her little boy somewhere.  It seemed to
be something that happened years ago. "  When asked if
she knew who it was, she said, "Somebody in the neighborhood where I live. "  When asked, she said it seemed

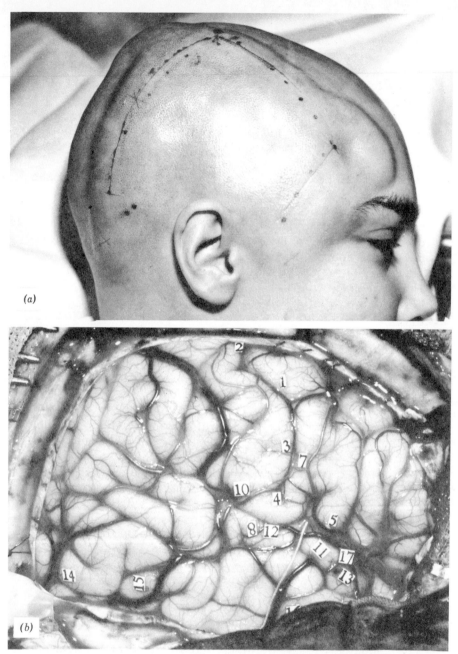

FIGURE 1-5.  (a) Preoperative view and (b) photograph of the cortex showing loci of stimulation for Case #36 described by Penfield and Perot (1963).

10

as though she was somewhere close enough to hear.

11. Repeated 18 minutes later. "Yes, I hear the same familiar sounds, it seems to be a woman calling. The same
lady. That was not in the neighbourhood. It seemed to
be at the lumber yard. "

13. "Yes, I heard voices down along the river somewhere--
a man's voice and a woman's voice, calling. " When
asked how she could tell it was down along the river,
she said, "I think I saw the river. " When asked what
river, she said, "I do not know, it seems to be one I
was visiting when I was a child. "

13. Repeated without warning. "Yes, I hear voices, it is
late at night, around the carnival somewhere--some sort
of a travelling circus. When asked what she saw, she
said, "I just saw lots of big wagons that they use to haul
animals in. "

12. Stimulation without warning. She said, "I seemed to
hear little voices then. The voices of people calling from
building to building somewhere. I do not know where it
is but it seems very familiar to me. I cannot see the
buildings now, but they seemed to be run-down buildings. "

14. "I heard voices. My whole body seemed to be moving
back and forth, particularly my head. "

14. Repeated. "I heard voices. "

The electrical exploration of M. M. 's cortex thus continues.
The number of these cases surveyed by Penfield and Perot (1963)
shows that such experiential response is not unique. They conclude that in the adult brain there is a remarkable record of the
stream of each individual's awareness or consciousness and that
stimulation of certain areas of cortex, lying on the temporal lobe
between the auditory sensory and the visual sensory areas,causes
previous experience to return to the mind of a conscious patient.
Penfield and Perot (1963) finally present clinicians, physiologists,
electronics experts, and psychologists with the challenging query
as to how these partially separable functional systems are integrated into normal brain activity and are related to the mind.

Needless to say, we "electronics experts" (at least) are a
long way from meeting this challenge. One response has been to
approximate the cortical neurons by several linear threshold units

as was suggested by McCulloch and Pitts in 1943.  It is then possible to proceed on to level 4 and to an appreciation of the reverberatory states and the temporal irreversibility exhibited by neural networks.  A key concept in relating neural network dynamics to the facts of psychology is that of the cell assembly introduced by Hebb in 1949.  In his view the cell assemblies are specially organized reverberations that constitute elements of thought, and an individual neuron may participate in many of them just as an individual member of society participates in many "social assemblies" (e. g. , citizen of Chicago, Sierra Club, Irish-American community, Catholic Church, Socialist party, neighborhood brass band, credit union, poker club, nuclear family, or clan).

The present author agrees with the substance of the strictures by Harmon and Lewis (1966) on the danger of irrelevancy that lurks for those who proceed on to level 4 without a sufficiently accurate representation of a real neuron.  Indeed, one of the main objectives of this book is to support current research efforts to understand and appreciate the "multiplex neuron. "  But we are also concerned with the tendency of physical scientists and behavioral psychologists to underestimate the functional complexity of the human brain.  In this context the "formal neuron" can be viewed as a conservative assumption, following which dynamic complexity is likely to be underestimated.

Since the number of cell assemblies in the human brain appear to be very large and their nature must differ substantially from person to person and from culture to culture, some fundamental difficulties can be expected to arise as scientists consider levels 6 and 7.  The traditional reductionist approach, which has proven itself so powerful in celestial mechanics, takes advantage of a simplicity and a determinism that are no longer available to those who deal with the dynamics of thought.

One need not retreat from natural philosophy to find mystery here, and those behaviorists who would reduce the human personality to a rational scheme should confront the full range of psychic phenomena.  To do less is not to be scientific but merely unobservant and therefore unrealistic.  Although there are some who would suggest that science is unrealistic (e. g. , Roszak, 1973), the present author supports attempts to broaden its purview and increase the flexibility of its inquiry so the charge cannot be sustained (e. g. , Maslow, 1966).

# 2

# Nonlinear Partial Differential Equations

*Qu.* 24. Is not Animal Motion perform'd by the Vibrations of this Medium, excited in the Brain by the power of the Will, and propagated from thence through the folid, pellucid and uniform Capillamenta of the Nerves into the Mufcles, for contracting and dilating them?

Newton

In this twenty-fourth question added to the second edition of his Optiks , Newton (1718) was fairly close to the mark, since a proper theory for nerve-fiber electrodynamics begins with the field equations of Maxwell, as does the science of optics. For a discussion on neurophysics it seems an appropriate place to begin. Paired with an understanding of membrane electrodynamics, the results of the present chapter provide a basis for the description of neuron behavior that the present author has designated as "level 3" in the scientific hierarchy.

Stimulated by attempts of Hermann and Matteucci to understand the manner in which electricity flows through a nerve fiber, Weber (1873a, b) carried out a fundamental study of time independent current density in and near a partially conducting cylinder. The basic coordinate system for this problem is shown in Fig. 2-1; a cylindrical membrane separates an inside region with conductivity

13

$\sigma_1$ and dielectric constant $\epsilon_1$ from an <u>outside</u> region with conductivity $\sigma_2$ and dielectric constant $\epsilon_2$. Weber, assuming that the electrical potential both inside and outside the membrane satisfies Laplace's equation, applied suitable boundary conditions at the membrane. This approach has been followed by several other investigators (Clark and Plonsey, 1966, 1968; Geselowitz, 1966, 1967; Hellerstein, 1968; Lorente de No, 1947; Plonsey, 1964, 1965; Rall, 1969; Weinberg, 1941, 1942) and indeed, is a very good approximation for potentials that vary as slowly as those indicated in Fig. 2-2. On the other hand, it is not more difficult to proceed with the complete Maxwell equations (Pickard, 1968, 1969; Rosenfalk, 1969; Scott, 1972), and this approach allows us to comprehend more precisely the implications of a quasistatic approximation. Thus we write

$$\text{curl } \overline{E}_i = -\mu_o \frac{\partial \overline{H}_i}{\partial t}$$

$$i = 1, 2 \qquad\qquad (2.1a, b)$$

$$\text{curl } \overline{H}_i = \sigma_i \overline{E}_i + \epsilon_i \frac{\partial \overline{E}_i}{\partial t}$$

where $i = 1$ inside the fiber and $i = 2$ outside. These equations are entirely <u>linear</u>. The nonlinearity in the problem appears at the membrane boundary where the normal current density, $J_{12}$, is some nonlinear function of the transverse voltage, v, across the membrane. We can write symbolically

$$J_{12} = N(v) \qquad\qquad (2.2)$$

but we must remember that $N(v)$ can be a rather complex function of v and its time derivatives. In order to appreciate this complexity, the reader might look ahead to the discussion of the Hodgkin-Huxley equations in Chapter 3.

The fact that nonlinear effects occur only on the cylindrical membrane boundary greatly simplifies the study of the electromagnetic problem. For an infinitely long fiber the regions both inside and outside the membrane are invariant to:

1.  translation in the x-direction;

2. rotation in the $\theta$-direction;
3. translation with time (t).

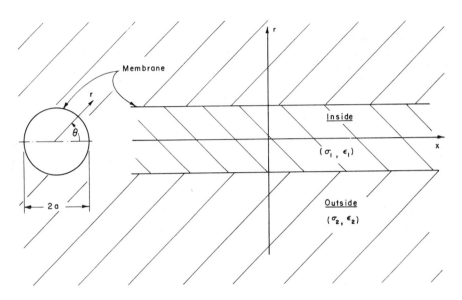

FIGURE 2-1.   Cylindrical geometry for electromagnetic analysis of a nerve fiber.

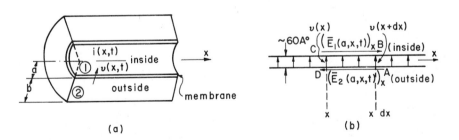

FIGURE 2-2.   (a) Geometry for an idealized nerve fiber;
(b) electric field components near the membrane.

Thus we can compose the fields of elementary functions that vary as $\exp[i(\beta x - \omega t + n\theta)]$ both inside and outside the fiber.

Furthermore, we begin our analysis by assuming rotational symmetry of the fields as implied by

$$\frac{\partial}{\partial\theta} = 0 \quad \text{or} \quad n = 0 \tag{2.3}$$

The implications of this assumption are considered in the text that follows, but for the present this assumption allows us to concentrate our attention on those TM (transverse magnetic) solutions of (2.1) for which*

$$(\overline{H})_r = (\overline{H})_x = 0 \quad \text{and} \quad (\overline{E})_\theta = 0 \tag{2.4}$$

We write the $\theta$-component of the magnetic intensity vector

$$(\overline{H})_\theta = H_\theta(r)\ \exp[i(\beta x + \omega t)] \tag{2.5}$$

and similarly for $(\overline{E})_r$ and $(\overline{E})_x$, where it sould be understood that subscripts 1 or 2 are added for fields inside or outside of the fiber. Maxwell's equations (2.1) then reduce to

$$\frac{\partial^2 H_\theta}{\partial r^2} + \frac{1}{r}\frac{\partial H_\theta}{\partial r} - \left(\frac{1}{r^2} + k^2\right)H_\theta = 0$$

$$E_r = -\frac{i\beta}{\sigma^*}H_\theta \tag{2.6a, b, c}$$

$$E_x = \frac{1}{\sigma^* r}\left[\frac{\partial(rH_\theta)}{\partial r}\right]$$

In these equations $\sigma^*$ is the complex conductivity

---

* The TE (transverse electric) modes for which $(\overline{E})_r = (\overline{E})_x = 0$ and $(\overline{H})_\theta = 0$ are of little interest, since the condition $(\overline{E})_r = 0$ implies zero normal current at the membrane surface. According to (2.2) such TE modes would not interact with the nonlinearity of the membrane.

$$\sigma^* = \sigma + i\omega\epsilon$$

and

$$k^2 = i\omega\mu_o\sigma^* + \beta^2 \qquad (2.\,6d, e)$$

Equations (2.6) indicate that $H_\theta$ is a rather convenient variable for which to solve. Knowing $H_\theta$, one can determine $E_r$ and $E_x$ through (2.6b) and (2.6c). Equation (2.6a) is Bessel's equation, solutions for which are $I_1(kr)$ and $K_1(kr)$ as defined by Watson (1962). Since $K_1$ goes to infinity at the origin, $I_1$ is the appropriate solution inside the membrane; and since $I_1$ goes to infinity for large values of $r$, $K_1$ is the appropriate solution outside. The magnitude of $H_\theta$ at $r = a$ can be easily determined using Ampere's circuital law [which is (2.1b) in integral form] from the total current flowing in the x-direction inside the membrane as

$$2\pi a H_\theta = I \qquad (2.7)$$

Thus a complete solution for H that: (a) satisfies Maxwell's equations both inside and outside the membrane, (b) has no $\theta$ variation as required by assumption (2.3), (c) corresponds to a TM mode with a current component perpendicular to the membrane boundary, (d) satisfies the appropriate electromagnetic boundary condition at the origin, and (e) goes to zero at large radius, is:

<u>inside</u>

$$H_\theta = \frac{I}{2\pi a} \frac{I_1(k_1 r)}{I_1(k_1 a)} \qquad (2.8a)$$

<u>outside</u>

$$H_\theta = \frac{I}{2\pi a} \frac{K_1(k_2 r)}{K_1(k_2 a)} \qquad (2.8b)$$

where $k_1^2 = i\omega\mu_o\sigma_1^* + \beta^2$ inside the fiber and $k_2^2 = i\omega\mu_o\sigma_2^* + \beta^2$ outside.

At this point in the analysis it is important to recognize that equations (2.8) have been derived without considering the nonlinear aspects of the problem symbolically expressed in (2.2). The appropriate values for $\omega$ and $\beta$ out of which the action potential is to be constructed are as yet entirely undetermined.   We now use (2. 8) to develop the nonlinear partial differential equations (PDE) that relate the total longitudinal current flowing inside the fiber, i(x, t), and the ($\theta$ independent) voltage across the membrane, v(x, t), as is indicated in Fig. 2-2a.

To obtain a PDE involving the x-derivative of v, consider the diagram of the electric field components near the membrane in Fig. 2-2b, where the positive reference directions for the x-component of the inside field, $(\overline{E}_1)_x$, and the outside field, $(\overline{E}_2)_x$, are indicated.   With these references, the sum of potentials around the path A → B → C → D becomes

$$v(x+dx) + [E_1(a, x, t)]_x dx - v(x) + [\overline{E}_2(a, x, t)]_x dx = 0$$

at any given time.   Thus

$$\frac{\partial v}{\partial x} = -[\overline{E}_1(a, x, t)]_x - [\overline{E}_2(a, x, t)]_x \qquad (2.9)$$

Equation (2. 9) is the source of the PDE we are after.   It can be related to the longitudinal current, i(x, t), in the following way. First consider the expansion of $(\overline{E}_1)_x$ into its spatial and temporal components as

$$[\overline{E}_1(a, x, t)]_x = \int\int E_{1x}(a)e^{i(\beta x + \omega t)}d\beta d\omega \qquad (2.10)$$

and similarly for $(\overline{E}_2)_x$.   Then using equations (2. 6c) and (2. 7) we can write

$$E_{1x}(a) = z_1 I \quad \text{and} \quad E_{2x}(a) = z_2 I \qquad (2.11a, b)$$

where $z_1$ and $z_2$ are impedances.   Assuming $H_\theta \to 0$ as $r \to \infty$ gives

$$z_1 = \left(\frac{1}{\pi\sigma_1^* a^2}\right)\left[\frac{k_1 a I_0(k_1 a)}{2I_1(k_1 a)}\right]$$

$$z_2 = \left(\frac{1}{\pi\sigma_2^* a^2}\right)\left[\frac{k_2 a K_0(k_2 a)}{2K_1(k_2 a)}\right] \qquad (2.12a, b)$$

Thus (2.9) becomes

$$\frac{\partial v}{\partial x} = - \int\int (z_1 + z_2) I(\beta, \omega) e^{i(\beta x + \omega t)} d\beta d\omega \qquad (2.13)$$

where $I(\beta, \omega)$ is the spatial and temporal Fourier transform of $i(x, t)$.

Equation (2.13) is not nearly as intractable in practice as it might appear at first glance. First, it is important to remember that it is entirely linear; the only nonlinearities appear in connection with current flow through the membrane (2.2), and this effect has not yet been considered. Second, the temporal frequency components, $\omega$, in a typical action potential are of the order of $10^3$ rad/sec (see Fig. 2-3) and the conductivity, $\sigma$, both inside and outside the fiber is approximately that of sea water (4 mho/m). Thus it is a very good approximation to write

$$\sigma^* \approx \sigma \quad \text{(a real constant)}$$

both inside and outside the fiber. Thirdly, the radial parameter k that appears in (2.12) is given by (2.6e)

$$k^2 = \frac{4i}{\delta^2} + \beta^2 \qquad (2.14)$$

where

$$\delta = \sqrt{\frac{2}{\sigma \mu_o \omega}} \qquad (2.15)$$

is the electromagnetic penetration depth in the conductive medium at frequency $\omega$. For $\sigma \sim 4$ mho/m, $\omega \sim 10^3$ radian/sec and $\mu_o = 4\pi \times 10^{-7}$ H/m, $\delta \sim 20$ m, which is much greater than the spatial extent of typical action potentials. Thus it is a very good approximation to write (2.14) as

$$k \approx \beta \qquad (2.16)$$

Finally, the spatial extent of a typical action potential is at least one order of magnitude greater than the fiber radius, a. Small argument approximations are then appropriate for the evaluation of the Bessel functions that appear in (2.12). For example,

(2. 12a) becomes

$$z_1 = \frac{1}{\pi a^2 \sigma_1} \{1 + 0[ (k_1 a)^2 ] \} \qquad (2.17)$$

and the most important effect of the $0[ (k_1 a)^2 ]$ terms is to introduce an inductive component into $z_1$. The effect of this inductive
component is studied in detail in Section 4-2, where it is shown
to be entirely negligible. Thus we can write

$$z_1 \approx r_1 \qquad (2.18)$$

where

$$r_1 = \frac{1}{\pi a^2 \sigma_1} \qquad (2.19)$$

is the resistance inside the fiber to longitudinal (x-directed) current flow. The ratio of outside to inside longitudinal impedance
from (2. 12) is

$$\frac{z_2}{z_1} = \left( \frac{\sigma_1}{\sigma_2} \right) \left[ \frac{K_0(\beta a) I_1(\beta a)}{K_1(\beta a) I_0(\beta a)} \right] \qquad (2.20)$$

where the square bracket is plotted in Fig. 2-3. Neglecting $z_2$
with respect to $z_1$ is seen to introduce an error of no more than
a percent.
    Thus (2. 13) reduces to

$$\frac{\partial v}{\partial x} = -r_s i \qquad (2.21a)$$

where $r_s = r_1 + r_2 \approx r_1$ as given by (2. 19). This is one of the two
partial differential equations we seek. The other is nonlinear and
relates the spatial derivative of $i$ to the membrane voltage $v$. It
is obtained from (2. 6b) by noting that $H_\theta$ is proportional to $i$ as
indicated in (2. 7) and, from (2. 6b), that $H_\theta$ evaluated at $r = a$
gives the current density normal to the membrane that appears in
(2. 4). Thus

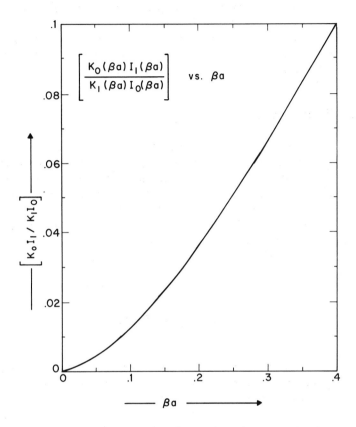

FIGURE 2-3.   Plot of the factor $(K_0I_1/K_1I_0)$   as a function of $\beta a$. This is approximately equal to the ratio of external to internal series resistance for

$$\frac{\partial i}{\partial x} = -2\pi a N(v) \qquad (2.21b)$$

It is often useful to combine equations (2.21) to obtain a second-order equation that involves only the membrane voltage

$$\frac{\partial^2 v}{\partial x^2} = 2\pi a r_s N(v) \qquad (2.22)$$

Equation (2.22) is not quite as simple as it appears, since $N(v)$ is a rather complex nonlinear function of $v$. But it is perhaps more simple than one expects for the geometry indicated in Fig. 2-2a. Thus it may be useful at this point to recapitulate the assumptions involved in the derivation of (2.22).

### a.  Rotation Symmetry of Fields

A basic assumption connected with (2.2) is that the membrane voltage is a function only of $x$ and $t$ and is not a function of the angle of rotation around the cylinder axis, $\theta$. In the course of the analysis, this restriction allowed us to exclude non-TM modes from consideration. Rall (1969) has studied the question of angle dependence in detail for a cylinder of fixed length. He has shown that the time constant, $\tau_n$, for angle variation [as $\exp(in\theta)$] to disappear is related to the basic time constant of the membrane, $\tau$, by

$$\frac{\tau_n}{\tau} \approx \frac{a}{n}\left(\frac{1}{\sigma_1} + \frac{1}{\sigma_2}\right)G \qquad (2.23)$$

where $G \equiv (\partial N/\partial v)$ is a conductance per unit area of the membrane. For typical values of the parameters, the right-hand side of (2.23) is something like $(10^{-4}/n)$. Thus for uniform cylindrical geometry, we can expect angularly dependent fields to relax to the angularly independent case in a time which is very short compared with the time scale for solutions of (2.22).

### b.  Uniform Fiber Cross-Section

A real nerve fiber is often not shaped as the uniform circular cylinder indicated in Fig. 2-2a; angular bends, local distention, tapering, and collapse into a ribbon shaped cross section are some of the deviations observed. Judgment is required to determine the degree of confidence that one can place in (2.22) in such cases. First, of course, the $\pi a^2$ that appears in (2.12) and (2.19) should be replaced by the cross-sectional area of the fiber, and the $2\pi a$ in (2.21b) and (2.22) should be replaced by the fiber circumference [Some calculations for flat cells were presented by Minor and

Maksimov (1969)]. A more serious difficulty arises from the scattering of TM fields [described by (2.22)] into non-TM modes; this effect is not represented at all. Furthermore, if the nonuniformities vary with x on a scale short with respect to $\beta^{-1}$ (the length of the action potential), the easy transition from (2.13) to (2.21a) will no longer be valid. On the other hand, some progress has been made with the solution of the nonlinear problem with a gradual taper, which is discussed in Section 4-6. Another important case is the so called "myelinated axon" for which $N(v)$ is approximately zero except at periodically spaced active nodes. This situation is also considered in detail in Section 4-5.

c.   Infinite External Medium.

In the development of the expression for outside impedance (2.12b) we assumed the dimension "b" in Fig. 2-2a sufficiently large to ensure that $K_1(k_2 b)$ was zero in (2.8b). From (2.16) a a more precise statement of this requirement is

$$b \gg \beta^{-1} \qquad (2.24)$$

where, as was noted before, $\beta^{-1}$ is of the order of the length of the action potential. Although this condition is easily satisfied in experiments on isolated fibers, it is also easily violated. Furthermore cells and fibers are often closely packed in functioning neural systems; thus the situation where (2.24) is not satisfied deserves attention.

If the external current is constrained to flow in a region $b \ll \beta^{-1}$ (i.e., very close to the membrane surface), the outside resistance increases from approximately zero to

$$r_2 \approx \frac{1}{A_o \sigma_2} \qquad (2.25)$$

where $A_o$ is the cross-sectional area outside the membrane. However, if $A_o$ does not exhibit rotational symmetry, the TM fields will again be scattered into non-TM modes in a manner not described by (2.22). Furthermore, if the changes in $A_o$ take place on a distance scale that is short relative to $\beta^{-1}$, the easy transition from (2.13) to (2.21a) will again no longer be valid. Qualitative effects of various experimental restrictions in the external geometry have been reviewed by Taylor (1963). Often nerve fibers are not isolated but arranged in bundles surrounded by a sheath of connective tissue.   The sciatic nerve of vertebrates (see Fig. 4-18) is constructed this way to permit the transmission of a multi-component message from the spinal cord to the muscle.   This situation has been investigated by Clark and Plonsey (1968), who present several numerical calculations which help to determine the effect of fiber geometry on $r_2$.

d.   Resistive Approximation for the Longitudinal Impedances.

    Equation (2. 22) specifically assumes that the sum of the inside and outside longitudinal impedances can be approximated by a single real number

$$r_s \equiv r_1 + r_2 \approx z_1 + z_2 \tag{2. 26}$$

This approximation ignores terms of order $(ka)^2$ in evaluating the small argument expressions for $z_1$ and $z_2$ in (2. 12).   Physically this implies neglect of the effect of time dependent magnetic fields on the electric field (i. e. , "inductive" effects).   In Section 4-2, following clarification of the nature of the nonlinear propagation process, we see that the only sensible effect of this inductive correction is to preclude a pulse velocity greater than the velocity of light.
    A transmission line equivalent circuit can easily be constructed to correspond to equations (2. 13) and (2. 21b).   For example, in the differential ladder network of Fig.  2-4a the change in series current over a differential distance, dx, is found from Kirchhoff's current law (or conservation of electric charge) to be

$$i(x) - i(x + dx) = 2\pi a \, N(v) dx \tag{2. 27}$$

which implies (2. 21b).   Similarly, the change in shunt voltage over a differential distance, dx, is obtained from Kirchhoff's voltage law ( or conservation of energy ) to be

$$v(x) - v(x + dx) = dx \; \mathscr{F}^{-1} \{ (z_1 + z_2) \mathscr{F} \, [\, i \,] \} \tag{2. 28}$$

where $\mathscr{F}$ and $\mathscr{F}^{-1}$, respectively, represent the Fourier transform on both x  and  t  defined as in (2. 10) and its inverse.   Equation

(2.28) implies (2.13).

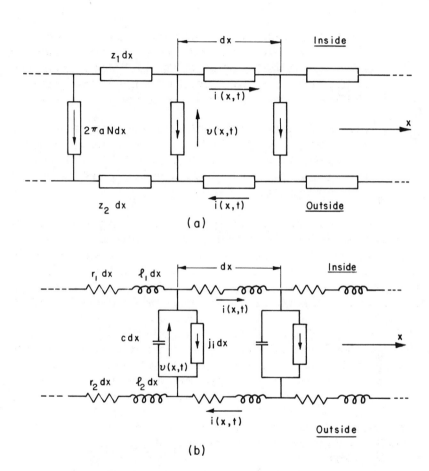

FIGURE 2-4.    (a)  Transmission-line equivalent representation of
(2.13) and (2.21b);  (b)  equivalent circuit for
a nerve fiber including series inductance.

Transmission line equivalent circuits (TLEC) of this sort have
found wide application in electronics since the development of the

electric telegraph (Kelvin, 1855) and in electrophysiology since
the turn of the century (Hoorweg, 1898; Hermann, 1905).  For
rather complete reviews, see Taylor (1963) and Cole (1968).  Various attitudes may be taken toward the TLEC, two of which are as
follows:

1.  The TLEC can be considered simply a mnemonic device
    through which the partial differential equations under con-
    sideration, namely, (2.13) and (2.21b), are represented
    pictorially.  It is often useful to suggest reasonable
    higher approximations for further study (Scott, 1970).

2.  The TLEC can be taken as the starting point for analysis.
    Equations (2.27) and (2.28) are then considered funda-
    mental equations from which (2.21b) and (2.13) are derived.
    This attitude has characterized much of past research in
    electrophysiology (Cole, 1968).

It is the opinion of the present author that the problems that
arise in studying the electrophysics of the nerve cell are sufficiently difficult that neither attitude should dominate.  For a
nerve fiber which approximates the idealized geometry of Fig. 2-2a,
it is clearly more satisfying (for the physicists, at least) to begin
the analysis with Maxwell's equations.  Various approximations
can be itemized and explicit analytic expressions can be obtained
for $z_1$ and $z_2$.  This analysis, on the other hand, can eventually
lead to the nonlinear PDE (2.22) that is also obtained directly from
(2.27) and (2.28).  In situations with more complex geometry,
where the electromagnetic analysis may not be tractable, one can
begin with a TLEC and appeal to the results for simpler geometry
as a justification.  Rall (1962, 1964) has demonstrated the power of
this approach through his application of "compartmental analysis"
to study the rather complex geometrical effects which arise in
dendritic fibers.

The general TLEC to be considered in the present work is
shown in Fig. 2-4b, for which suitable expressions to determine
$r_1$ and $r_2$ are given in (2.19) and (2.25).  With the series inductances ($\ell_1$ and $\ell_2$) equal to zero, this TLEC was studied by
Offner as early as 1937 and serves as the basis for the calculation
of conduction velocity for an action potential by Offner, Weinberg,
and Young (1940).  We continue to assume these inductances equal
to zero for the initial development of the nonlinear analysis.  In a

later chapter explicit expressions and values are calculated, and a nonlinear propagation problem is solved in order to demonstrate that is is a valid assumption to take these inductances equal to zero. Notice that the shunt element in Fig. 2-4b is represented differentially than that in Fig. 2-4a. The reason for this change is that in Fig. 2-4b it is explicitly recognized that membrane current consists of two distinct components; namely, displacement current and ion current. Equating the shunt currents in the two figures yields

$$2\pi aN = c \frac{\partial v}{\partial t} + j_i \qquad (2.29)$$

where $j_i$ is the ion current and $c(\partial v/\partial t)$ the displacement current passing through the membrane, both per unit length in the x-direction. The decomposition indicated in (2.29) is especially interesting because there is substantial experimental evidence (Cole, 1968) to show that c is a constant throughout the course of the action potential [see, however, FitzHugh and Cole, 1973). Substituting (2.29) into (2.22) yields a new form for the basic equation of nerve propagation

$$\boxed{\frac{\partial^2 v}{\partial x^2} - r_s c \frac{\partial v}{\partial t} = r_s j_i} \qquad (2.30)$$

Notice that (2.30) has the form of a nonlinear diffusion equation. In Chapter 3 we consider the chemical physics of the nerve membrane and the development of phenomenological theories to describe the nonlinear dependence of $j_i$ on v.

# 3

# The Active Nerve
# Membrane

To see a World in a Grain of Sand
And a Heaven in a Wild Flower,
Hold Infinity in the palm of your hand
And Eternity in an hour.

Blake

In this chapter we consider what is currently known about
level 2 of the scientific hierarchy sketched in Chapter 1. Although
there are several useful phenomenological representations of the
nerve-membrane electrodynamics, the present state of knowledge
is frustrating and unsatisfactory because a widely acceptable explanation for the phenomena has not yet emerged. However, the
present author expects this gap to be filled within the next few
years.

## 1. PHYSICS OF A CELL MEMBRANE

Our first task is to become acquainted with the physical character of the cell membrane, which is indicated merely as a surface
in Fig. 2-1 and as a homogeneous region in Fig. 2-2. The existence of a membrane for red blood cells was confirmed by the
measurements of Fricke (1923, 1925a, b, 1926) on the conductivity,
plotted against frequency, of cell suspensions. He measured a
membrane capacitance of 0.81 $\mu F/cm^2$, which for an assumed relative dielectric constant of 5, implied a membrane thickness of
55Å. At about the same time Gorter and Grendel (1925) demonstrated

that these cells "are covered by a layer of fatty substances that is two molecules thick. " It is well to devote a moment to the measurement technique of Gorter and Grendel because it nicely exemplifies the energetics of membrane structures. The general structure of a lipid (fatty) molecule is "cigar-shaped" with a charged head group localized at one end of a hydrocarbon tail. [ See Chapter 10 of Lehninger (1970) for many chemical details. ] Building on a previous demonstration by Lord Rayleigh that oil films on a water surface become monomolecular, Langmuir (1917) (see also Adam, 1921, 1922) showed that the structure of the mono-layer is with the charged head groups oriented toward the water surface where the electric field energy can be reduced by the high dielectric constant of water ($\sim 80 \ \epsilon_o$), and the hydrocarbon tails maintained in a closely packed, vertical structure by transverse Van der Waals attraction. Gorter and Grendel distilled the lipid material from a known quantity of blood cells and found the area of the monolayer obtainable with this lipid at an air-water inter-face to be about twice the area of the cell surfaces. Thus the red blood cell membrane appeared to be largely the lipid bilayer shown in Fig. 3-1. This same structure was proposed (Danielli and Dav-son, 1935 ; Danielli, 1936) from an energetic comparison of various lipid organizations, as the basic structure of biological cell mem-branes. Membrane distillates always contain a substantial frac-tion (> 50%) of protein (Bretscher, 1973; Kilkson, 1969); and if these are located within the lipid phase they are called intrinsic (Green, 1971) or integral (Singer and Nicolson, 1972). Proteins attached weakly to the surface of the lipid bilayer, called extrin-sic or peripheral, are considered to be of less importance for mem-brane function. Green, Ji and Brucker (1972) have emphasized the importance of protein domains through which long-range ordering of (perhaps octal) protein subunits is established (Vanderkooi and Green, 1970) as indicated in Fig. 3-1. Singer and Nicolson, on the other hand, have suggested that the proteins may be consider-ed to float in the two-dimensional lipid liquid. Good general sur-veys of biological membranes are given in the recent books by Cereijido and Rotunno (1970), Jain (1972), and Nystrom (1973), and many of the historically important papers have been collected by Branton and Park (1968). See Markin and Chizmadzhev (1974) for a recent survey of the Russian literature. The direct synthesis of a biological membrane was attempted by Bundenberg de Jong and Bonner (1935), Devaux (1936), Danielli (1936), Teorell (1936),

Langmuir and Waugh (1938), and Dean (1939), who produced bilayer films with a capactiance of about 1 $\mu F/cm^2$ (Dean, Curtis,
and Cole, 1940).  This work lay dormant for more than two decades
until Mueller, Rudin, Tien, et al. (1962) demonstrated the ease
with which lipid bilayers can be formed.  The key idea was an observation in Newton's Optiks on the color patterns of soap bubbles.
He had observed that "after all the Colours were emerged at the
top, there grew in the center of the Rings a small round black
Spot . . . which continually dilated itself till it became sometimes more than 1/2 or 3/4 of an inch in breadth. " Newton observed that it is energetically favorable for a soap film to thin into a
lipid bilayer.  In this case the charged head groups are oriented
inward toward a remnant layer of water.  Such a soap film appears
"black" (i. e. , almost reflectionless) because its thickness
($\sim 100$ Å) is very much less than the wavelength of light.  [ The
reader who wishes to learn more about this subject should consult
the delightful descriptions prepared by Lawrence (1929) and
Mysels, Shinoda, and Frankel (1959). ]  Mueller, Rudin, Tien, et
al. (1962) showed that the same result could be obtained for lipid
films between aqueous phases.

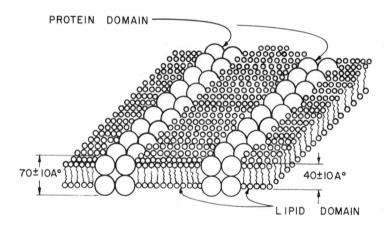

FIGURE 3-1.  The "structure-function unitization model"(redrawn
from Green, Ji, and Bruckner , 1972). Domain geometry
is assumed to be highly variable from membrane to
membrane.

A diagram of the basic arrangement for measurements on artificial lipid bilayers is given in Fig. 3-2.  A camel's hair brush is
dipped in the lipid solution and then stroked across a small ($\sim 1$ mm)
hole in a two-chamber vessel.  The resulting thick lipid film thins
in about 10 min, as Newton described, to the lipid bilayer of black
film.  Optical measurements of film thickness and electrical measurements of capacity and conductivity can then proceed.  The ex-
perimentalist who wishes to begin such an investigation is referred to the review by Goldup, Ohki, and Danielli (1970), the careful discussion of experimental details by Howard and Burton (1968),
and the recent book by Jain (1972).

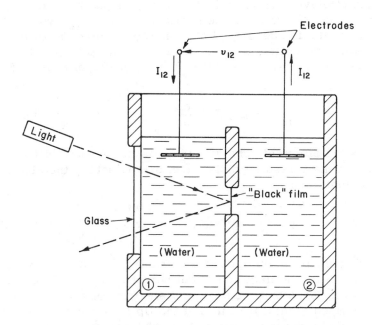

FIGURE 3-2.  Basic experimental arrangement for physical
             measurements on artificial lipid bilayers.

The processes by which ions flow across the membranes of
living cells are often classified as either _passive_ and _active_ mechanisms.  Passive transport is considered to be in response to a

gradient of the electrochemical potential.    Active transport involves the flow of ions against the electrochemical potential; a
good discussion of such processes is in Chapter 27 of Lehninger
(1970).    During the propagation of an action potential along a
nerve fiber (Fig. 1-3) only passive transport is involved; active
processes merely recharge the energy sources.    My objective here
is to present a simple phenomenological description of passive
transport from which the ionic current components in (2. 2) can be
constructed.

It should be understood from the start that intrinsic membrane
proteins completely dominate ion flow in a living membrane.    To
appreciate the truth of this assertion, it is instructive to begin with
an investigation of passive transport of only sodium ions across
an ideal lipid bilayer, as is indicated in Fig. 3-3.    The steady-
state current density from chambers    ①    to    ②  , $J_{12}$ , will be
proportional to the ion density [ $Na^+$]    and to the gradient of the
electrochemical potential, $\psi$ .    Thus we can write the Nernst-
Planck equation (Nernst, 1888, 1889; Planck, 1890a, b; Smith, 1961)

$$J_{12} = - \mu q [ Na^+ ] \frac{d\psi}{dr} \tag{3.1.1}$$

where q  is the electronic charge and  $\mu$  is the ionic mobility.
Assuming that the pressure gradient can be neglected,  the electrochemical potential is

$$\psi = \frac{kT}{q} \log [ Na^+ ] + v + w \tag{3.1.2}$$

where v  is the externally applied electrical potential and  w  is
the contribution to the electrochemical potential arising from the
presence of the membrane.    Since the dielectric constant of water
($\sim 80 \ \epsilon_o$) is much greater than that of the lipid, a major contribution to  w  will be the image force at the water-membrane boundary.    (In direct physical terms, the electrostatic field energy associated with the ion is much lower in the water phase where the
ionic charge can be neutralized by rotating water molecules. ) The
factor (kT/q) in (3. 1. 2) (where  k  is the Boltzmann constant and
T  is absolute temperature) appears from the Einstein (1905) relation between diffusion constant and mobility.

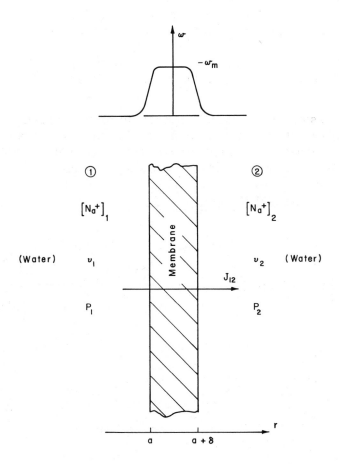

FIGURE 3-3.    Simple geometry for passive transport of a single ion through a uniform membrane.

Substituting (3. 2) into (3. 1) gives

$$J_{12} = -\mu KT \{ \frac{d[Na^+]}{dr} + \frac{[Na^+]q}{kT} \frac{d}{dr}(v+w) \} \qquad (3. 1. 3)$$

which, in steady state, must be independent of $r$.  The boundary conditions to be satisfied at the edges of the membrane are

$$r = a: \qquad [Na^+] = [Na^+]_1, \qquad v = v_1, \qquad w = 0$$

$$\text{(3.1.4a, b)}$$

$$r = a + \delta: \quad [Na^+] = [Na^+]_2, \qquad v = v_2, \qquad w = 0$$

The expression for sodium-ion concentration that satisfies these boundary conditions and maintains $J_{12}$ constant has been determined by Neumke and Läuger (1969) as (see also Boltaks, Vodyanoi, and Fedorovich, 1971 and Markin and Chizmadzhev, 1974):

$$[N_a^+] = \exp\frac{-(v+w)q}{kT}\left\{ [N_a^+]_1 \exp\frac{v_1 q}{kT} \right.$$

$$+ \left[ [N_a^+]_2 \exp\frac{v_2 q}{kT} - [N_a^+]_1 \exp\frac{v_1 q}{kT} \right] \frac{\displaystyle\int_a^{a+r} \exp\frac{(v+w)q}{kT} dr}{\displaystyle\int_a^{a+\delta} \exp\frac{(v+w)q}{kT} dr} \left.\right\}$$

$$\text{(3.1.5)}$$

which on substitution into (3.1.4) becomes

$$J_{12} = \frac{\mu kT}{\displaystyle\int_a^{a+\delta} \exp\frac{(v+w)q}{kT}dr} \left\{ [N_a^+]_1 \exp\frac{v_1 q}{kT} - [N_a^+]_2 \exp\frac{v_2 q}{kT} \right\}$$

$$\text{(3.1.6)}$$

For a detailed discussion of the effect of barrier shape, $w(r)$, on volt-ampere characteristics, see Hall, Mead, and Szabo (1973). Under experimental conditions for which the membrane structure remains independent of the applied voltage, those authors have demonstrated that $w(r)$ can be computed from volt-ampere measurements. The measured barrier height closely approximates the difference in electrostatic energy of the ion in lipid and in water; the shape is trapezoidal, as indicated in Fig. 3-3.

Introducing the notational definitions

$$v_{12} \equiv v_1 - v_2 \tag{3.1.7}$$

and

$$V = \frac{kT}{q} \log \frac{[N_a^+]_2}{[N_a^+]_1} \tag{3.1.8}$$

we obtain from (3.1.6)

$$J_{12} = \frac{kT}{q} G \left[ \exp \frac{(v_{12}-V)q}{kT} - 1 \right] \tag{3.1.9}$$

$$\approx G(v_{12} - V) \quad \text{for} \quad |v_{12}-V| < \frac{kT}{q} \tag{3.1.10}$$

where

$$G \equiv \frac{\mu q [N_a^+]_2 \, \exp \dfrac{v_2 q}{kT}}{\displaystyle\int_a^{a+\delta} \exp \dfrac{(v+w)q}{kT} \, dr} \tag{3.1.11}$$

From (3.1.2) it is seen that $(v_{12} - V)$ is the change in electro-chemical potential from chambers ② to ① . For a sufficiently small difference in electrochemical potential, (3.1.10) indicates that the relation between voltage and ion-current density should be linear. If the concentration gradient is zero, $[N_a^+]_1 = [N_a^+]_2$, this linear relation should go through the origin as indicated in Fig. 3-4a. If $[N_a^+]_1 < [N_a^+]_2$ and the current density is zero, chamber ① will have a positive potential with respect to ② . If $[N_a^+]_1 > [N_a^+]_2$, the polarity of the zero current voltage difference will be reversed. Thus each ionic species appears in the membrane as (see Fig. 3-4b) a battery of voltage given by (3.1.8) with its positive terminal directed toward decreasing (increasing) ion concentration for positive (negative) ions. In general there will be several species of ions present which makes the analysis considerably more difficult. In 1943 Goldman derived a generalization of (3.1.6) under the assumption of a constant electric field (electroneutrality), and Offner (1971) has recently discussed numerical techniques that do not require this assumption. See Rosenberg (1969) for a recent comparison of resting-potential formulas.

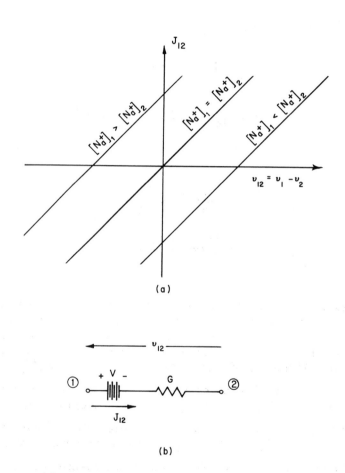

(a)

(b)

FIGURE 3-4.    (a)  Sodium ion current density at a small difference of electrochemical potential; (b) equivalent circuit for the current density carries by sodium ions.

Let us now consider an experiment in which a pure lipid bilayer is carefully prepared in the apparatus of Fig. 3-2 (Howard and Burton, 1968) with equal concentrations for all ions so the ionic batteries are all zero. The initial slope of the current density-voltage curve can be as low as (Goldup, Ohki, and Danielli, 1970)

$$G \sim 10^{-9} \text{ mho/cm}^2$$

which, for a membrane thickness of about 100 Å ($\sim 10^{-6}$cm), implies a membrane resistivity

$$\rho \sim 10^{15} \text{ ohm-cm}$$

Thus a clean lipid bilayer should be classified as a very good insulator, and the importance of the protein complex in facilitating ionic conduction through biological membranes cannot be overemphasized (Ehrenstein, 1976).

Mueller, Rudin, Tien, et al. (1962) showed that the addition of small amounts of properly chosen and refined proteinaceous material (called EIM for "excitability inducing material") will increase the membrane conductivity by many orders of magnitude and can introduce the nonlinearity essential for generation of an action potential. At low protein concentrations the conductance has been observed to increase in quantum units of about $4 \times 10^{10}$ mhos (Goldup, Ohki, and Danielli, 1970). When alamethicin (a circular polypeptide with molecular weight $\sim 1800$) is added to the aqueous phase of a clean experiment, the membrane conductance is found to increase with the sixth power of concentration (Mueller and Rudin, 1968b). These observations suggest that the alamethicin molecules may be coordinating in groups of six to permit ionic conduction through the membrane. Hille (1970) has surveyed a wide variety of kinetic, electrochemical, and pharmacological data for biological nerve membranes and has concluded that the conductance changes observed during the action potential (see Fig. 1-3) are caused by the opening and closing of localized conductance channels. The term "pore" is often used in a generic sense to indicate a localized region of high conductivity on the membrane. For such a porous membrane (3.1.6) is no longer useful. The barrier potential, $w(r)$, and the ionic mobility, $\mu$, depend strongly on the position on the membrane surface and also on the membrane voltage.

In this situation it is helpful to return to (3.1.1) and write it in the form

$$J_{12} = G(v_{12} - V) \qquad (3.1.12)$$

where, as was previously noted, $(v_{12} - V)$ is the negative of the change in electrochemical potential from chambers ① to ② . The conductivity is not a constant but a nonlinear function of the experimental variables. The form of (3.1.12) merely makes explicitly evident the zero in ion current that appears when the electrochemical difference for that ion is zero. The conductance per unit area, $G$, often appears as a function only of the transmembrane voltage, $v_{12}$. An exceptionally clear example of this has recently been presented by Eisenberg, Hall, and Mead (1973) in connection with their careful study of the effect of alamethicin on artificial lipid bilayer membranes. The volt-ampere curve in Fig. 3-5a exhibits a distinct region of negative differential conductance, but the conductance (see in Fig. 3-5b) shows a simple exponential rise throughout this region. The experimental rise is the same as that observed without an ion imbalance. Thus it is clear that in this case we can write (3.1.12) in the form

$$J_{12} = G(v_{12})(v_{12} - V) \qquad (3.1.13)$$

As has been pointed out by Cole (1968, p. 289) and by Mueller and Rudin (1968a, b), the condition for negative differential conductance can then be expressed by differentiating (3.1.13) with respect to $v_{12}$.

$$\frac{dJ_{12}}{dv_{12}} = G'(v_{12} - V) + G$$

so

$$\frac{dJ_{12}}{dv_{12}} < 0 \Rightarrow G'(V - v_{12}) > G \qquad (3.1.14)$$

This condition for negative differential conductance was first demonstrated for an alamethicin doped artificial lipid bilayer membrane by Mueller and Rudin (1968b). Whenever membrane current

($J_{12}$) is related to membrane voltage ($v_{12}$) as in (3. 1. 13), the condition can be expressed in the following simple physical terms:
negative differential conductance will appear when $G$ is rising
rapidly enough below the resting voltage.   Since the resting voltage depends on ion concentrations, negative differential conductance of a membrane can be made to appear or disappear simply by
changing the composition of the external solutions!   Thus as Agin
(1969) has emphasized, the mere appearance of a negative conductance need not depend on exotic effects such as interaction of
divalent ions, conformational changes of macromolecules, micelle
transformations of lipid systems, enzyme reactions, ion-specific
carriers, redistributions of pores, and chemical gates.

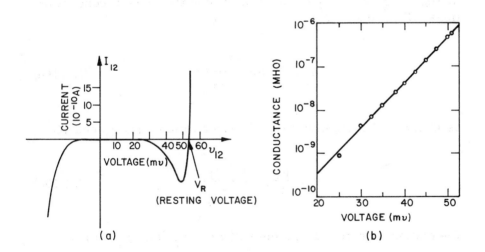

FIGURE 3-5.    Measurements on an artificial lipid bilayer membrane
in a 100:1 KCl gradient.   Chamber ①, 0. 005 m KCl
and $9 \times 10^{-6}$ g/ml alamethicin;   Chamber ②,
0. 5 m KCl  and  $6 \times 10^{-7}$ g/ml alamethicin.   From
Eisenberg, Hall and Mead (1973).   (a) Current
plotted against voltage; (b)  conductance plotted
against voltage.

Cole (1968, pp. 287-290) points out that the functional form in (3.1.12) is especially useful for description of a squid-axon membrane since G remains constant for times up to the order of 100 μsec. The current flow in response to more rapid changes in voltage is simply ohmic.

It should be noted that the current indicated in Fig. 3-5a is due to <u>both</u> potassium and chlorine ions. In general a membrane which separates n-ionic species can be represented as in Fig. 3-6, where current is related to transmembrane potential by (Cole, 1968, pp. 193-197)

$$J_{12} = [\sum_{i=1}^{n} G_i] v_{12} - \sum_{i=1}^{n} (G_i V_i) \qquad (3.1.15)$$

which has the same form as (3.1.12). From the discussion related to Fig. 3-4 it should be clear that for positive ions of concentrations $[C^+]_1$ and $[C^+]_2$

$$V_i = \frac{kT}{q} \log \frac{[C^+]_2}{[C^+]_1} \qquad (3.1.16a)$$

as (3.1.8). For negative ions of concentrations $[C^-]_1$ and $[C^-]_2$

$$V_i = \frac{kT}{q} \log \frac{[C^-]_1}{[C^-]_2} \qquad (3.1.16b)$$

The resting potential (i.e., the value of $v_{12}$ for $J_{12} = 0$) is

$$V_R = \frac{\sum (G_i V_i)}{\sum G_i} \qquad (3.1.17)$$

Thus if the conductance, G, for a particular ion becomes large, the resting potential will approach the battery voltage for that ion. To see how these equations can be used, consider the data in

Fig. 3-5a.   The resting potential, $V_R$ = 53 mV  and, from the ion
concentration ratios and (3.1.16), $V_K$ = +115 mV  and  $V_{Cl}$ = -115mV.
Thus from (3.1.17) we find at the resting potential that ($G_K/G_{Cl}$ =
2.7 so about 73% of the ion current flowing in the vicinity of the
resting potential should be carried by potassium ions.

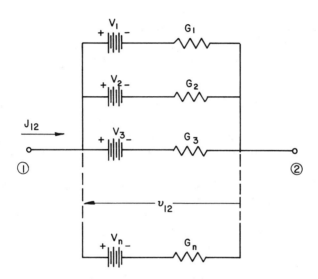

FIGURE 3-6.   Membrane equivalent circuit for  n  ionic species.


Depending on one's point of view,  (3.1.15) can be considered
either:  (a)  a flexible and useful description of multicomponent
ion flow or (b)  a phenomenological representation without physical
meaning.   The second attitude has been presented in detail by
Tasaki (1968), who points out that if no restrictions are placed on
the functional dependence of the  $G_i$'s ,  then  (3.1.15) says nothing more than (3.1.12).   Furthermore (3.1.17) is of no value for
calculation of a resting potential unless other information about
the membrane permeability to various ions is available.   Tasaki
(1968) carefully considers the calculation of resting potentials from
physical considerations under a variety of simplifying assumptions.

A complementary discussion is given in the recent book by Khodorov (1974).

As an example of the kind of equation that can be derived for the resting voltage, Hodgkin and Katz (1949) assumed that each ion obeys the Nernst-Planck equation (3.1.3) and that the ith ion concentration just inside the membrane is a partition coefficient, $\gamma_i$, times the corresponding concentration outside the membrane. Then for univalent ions

$$V_R = \frac{kT}{q} \log \left( \frac{\sum^+ \mu_i \gamma_i [\, C^+\,]_2 + \sum^- \mu_i \gamma_i [\, C^-\,]_1}{\sum^+ \mu_i \gamma_i [\, C^+\,]_1 + \sum^- \mu_i \gamma_i [\, C^-\,]_2} \right) \qquad (3.1.18)$$

where $\sum^+ \left( \sum^- \right)$ indicates summation over the positive (negative) ions.

Thus far we have been considering only passive (i. e. , non-metabolic) mechanisms for ion transport across a cell membrane. Active ion transport is extremely important in the operation of a living cell, and, although the details of such processes are not yet well understood, the broad outlines are emerging (Lehninger, 1970). The inside of a nerve cell, for example, is usually some 60-70 mV negative with respect to the outside. Using the convention of Figs. 2-2 and 3-3.

$$J_{12} = 0 \quad \text{for} \quad v_{12} = V_R \approx -65\,mV$$

For the squid giant axon (Hodgkin and Huxley, 1952)

$$\frac{[\, Na^+\,]_2}{[\, Na^+\,]_1} \approx 7.5 \Rightarrow V_{Na} = +50\ mV$$

and

$$\frac{[\, K^+\,]_1}{[\, K^+\,]_2} \approx 30 \Rightarrow -85\,mV$$

Thus metabolic energy must be expended to pump sodium ions outward against the resting potential. We shall see in Chapter 4 that the electric field energy associated with the resting potential is expended in the propagation of an action potential (Hodgkin, 1964). Current knowledge of the processes for outward pumping of $Na^+$ and inward pumping of $K^+$ has recently been reviewed by Thomas (1972). There is indication that three sodium ions are removed for each two potassium ions that enter. The energy for this process is supplied by the conversion of ATP (adenosinetriphosphate) to ADP (di). The ATP, in turn, is reconstituted in the membranes of subcellular units known as mitochondria.

## 2. ELECTRODYNAMICS OF AN ACTIVE NERVE MEMBRANE

The most extensive nerve membrane measurements have been performed on the giant axon of the squid [ see Cole (1968) for a thorough discussion of the literature and a beautiful color photograph of the animal]. This fiber is between 0. 5mm and 1 mm in diameter and several centimeters in length. It is easily removed from the squid and continues to function for at least several hours and often as long as a day. *

A typical experimental arrangement for measuring the electrodynamic properties of a membrane is indicated in Fig. 3-7 (Hodgkin, Huxley, and Katz, 1952). This is called a "space-clamped" measurement because the electrode arrangement eliminates the possibility of longitudinal variation of voltage and the associated wave propagation effects; it is also called a "voltage-clamped" measurement if a negative feedback amplifier is introduced to reduce the source impedance and permit $v_{12}$ to be independently specified. We are interested in interpreting the relationship between $J_{12}$ and $v_{12}$ to extract the nonlinear character of the membrane indicated simply by $J_{12} = N(v_{12})$ in (2. 2). As was previously mentioned in connection with (2. 29), $J_{12}$ is

---

* An introduction to the surgical procedures for removal of a nerve is provided by the two-part film loop, Nerve Impulse, available from Ealing Corp. , 2225 Massachusetts Avenue, Cambridge, Mass. 02140. Arnold, Summers, Gilbert, et al. (1974) give many useful hints for those who would experiment with the squid giant axon.

composed of a displacement current component through the membrane capacity and an ion current through the membrane. Thus

$$J_{12} = C\frac{dv_{12}}{dt} + J_i \qquad (3.2.1)$$

where C is the capacitance per unit area of the membrane ($\sim 1\ \mu F/cm^2$) and $J_i$ is the sum of all the individual ionic currents through the membrane. If $v_{12}$ is independent of time, the displacement current is zero, and the ion current should be a sum of terms as in (3.1.15). Therefore, in measuring the ionic currents it is convenient to hold the membrane voltage fixed. It was this "voltage-clamp" measurement (Cole, 1949, Marmont, 1949, Ramon, Moore, Joyner, et al., 1975) that led Hodgkin and Huxley (1952) to a representation of $J_i$ that could be used to solve (2.30) for a propagating action potential.

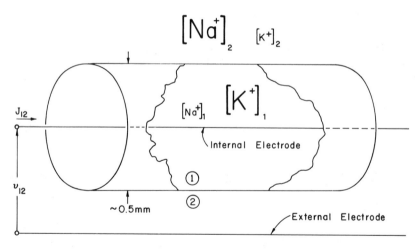

FIGURE 3-7.    Geometry for a space-clamped measurement on a squid nerve membrane.

The sodium and potassium ion currents are most interesting because they respond nonlinearly to changes of voltage across the membrane. The behavior of these nonlinear currents has been described in a simple and appealing way by Katz (1966) and by Cole (1968) using the equivalent circuit shown in Fig. 3-8a. This representation includes a sodium battery of about 50 mV directed inward and a potassium battery of about 77 mV directed outward. As was noted in the previous section, the ion batteries account for the tendency of sodium ions to diffuse inward and for potassium ions to diffuse outward. These batteries are in series with a sodium conductance per unit area, $G_{Na}$, and a potassium conductance per unit area, $G_K$, respectively, as is indicated in Fig. 3-6. A small boy (named "Nat") senses the voltage across the membrane and adjusts $G_{Na}$ according to some rules of his own, and another small boy (named "Kal") does the same for $G_K$. What Nat and Kal do is conveniently described in terms of the change of potential inside the membrane with respect to its resting value. Thus we define

$$v \equiv v_{12} - V_R \qquad\qquad (3.2.2)$$

If the voltage inside the membrane is rendered more negative (hyperpolarized), the membrane conductances remain small with little change in value. If the voltage inside the membrane is made less negative (depolarized) the reactions of Nat and Kal are indicated in Fig. 3-8b, c (Hodgkin and Huxley, 1952d). The individual ion-current components can be measured by assuming the validity of (3.1.15) and adjusting the external salt solution to make $(v_{12} - v_{Na})$ or $(v_{12} - V_K)$ equal to zero at some particular value of $v_{12}$.

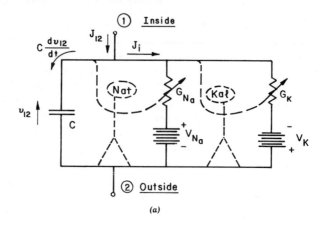

(a)

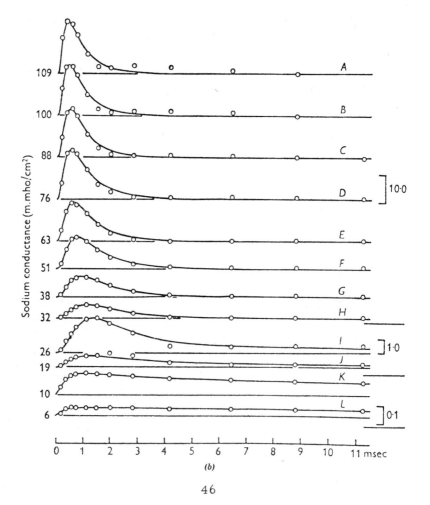

(b)

46

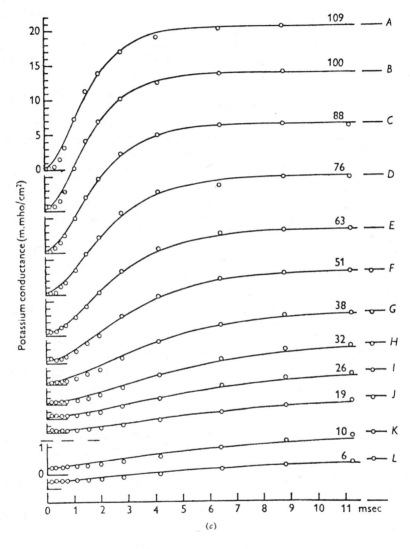

FIGURE 3-8.  (a)  A simplified equivalent circuit for a unit area of squid membrane; (b)  reaction of small boy "Nat" to displacement of membrane voltage from the resting value; (c) ditto for "Kal" (Hodgkin and Huxley, 1952d).

47

The curves in Fig. 3-8 indicate the way $G_{Na}$ and $G_K$ change with time for a <u>fixed</u> change in voltage. If the circuit is not volt-age clamped, however, it will "switch." The reason for this is that a small depolarizing voltage ($v > 25$ mV) increases the con-ductance of the membrane to sodium ions. Thus sodium ions flow <u>into</u> the membrane, which <u>increases</u> the depolarizing voltage caus-ing the sodium ion conductance to increase even further. It is a positive feedback effect; once initiated the membrane will rapidly approach the sodium ion battery voltage ($v_{12} = v_{Na}$ or $v = V_{Na} - V_R = 115$ mV) due to the rapid inflow of sodium ions. Then $G_{Na}$ will fall back toward zero (Fig. 3-8b) and $G_K$ will rise (Fig. 3-8c), allowing an outflow of potassium ions. This outward potassium ion current will bring the membrane potential back to its resting value. Increasing the potential inside the membrane by 25 mV or more is something like pulling the chain on a hopper; once the pro-cess starts it goes through a complete cycle. In large fibers the total ionic flow during one switching cycle is a very small fraction of the total ion concentration; many hundreds of thousands of fir-ings can occur in a squid giant axon before the ionic batteries be-come discharged. In smaller fibers, such as those shown in Fig. 1-2, the ionic flow per impulse can be a substantial fraction of the total ion concentration.

This is a description of <u>what</u> happens. <u>Why</u> it happens is not yet understood, but some interesting clues can be gleaned from an investigation of the total ion current that flows in response to a fixed voltage (so $J_{12} = J_i$). From Fig. 3-9 it can be seen that if the voltage $v_{12}$ is held at a value less than $V_{Na}$, the current $J_{12}$ is first negative (inward) and then positive (outward). From these curves it is possible to define an initial peak, $J_p$, and a final steady-state value, $J_{ss}$, as is indicated for the curve at $v_{12} = -20$ mV in Fig. 3-9. Both $J_p$ and $J_{ss}$ can then be plotted against the corresponding value of the voltage step as is indicated in Fig. 3-10a. The early, $J_p$, branch of the curve is primarily sodium-ion current; while the steady state, $J_{ss}$, branch is primar-ily potassium-ion current. The membrane appears to be in a high conductance state for $v_{12} > -40$ mV and a low conductance state for $v_{12} < -50$ mV. Returning to the inequality condition for differ-ential negative conductance expressed in (3.1.14), we see that in the range $-50$ mV $< v_{12} < -40$ mV the conductance is rising "rap-idly enough."

Similar data for other electrically active biological membranes

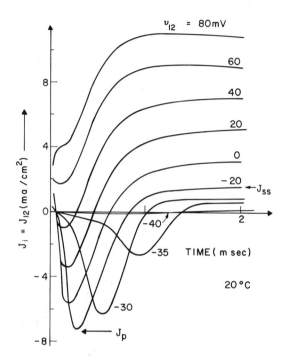

FIGURE 3-9.   Typical response of squid membrane current density
to fixed steps of voltage (Cole, 1968, p. 326).
Published in 1968 by The Regents Of The University
Of California; reprinted by permission of the
University of California Press.

are plotted in Figs. 3-10b-f.  In each case there is an early current
density $(J_p)$ or current $(I_p)$ exhibiting negative differential conductivity and eventually relaxing into a steady-state current density $(J_{ss})$ or current $(I_{ss})$ with only positive differential conductivity.  In Figs. 3-10a-d the early current is carried primarily by
sodium ions and the later current is carried primarily by potassium

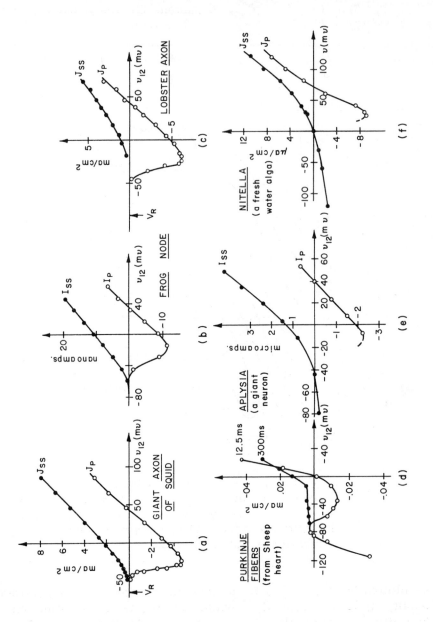

FIGURE 3-10. Voltage-clamp data from various active biological membranes: (a), (b), and (c) redrawn from Cole (1968); (d) redrawn from Deck and Trautwein (1964); (e) redrawn from Guduldig and Gruener (1970); (f) redrawn from Kishimoto (1965). $J_p$ and $J_{ss}$ are defined as in Fig. 3-9.

50

ions.  In the measurement on membranes from <u>Aplysia californica</u> shown in Fig. 3-10e, Geduldig and Gruener (1970) find clear evidence for a calcium ion contribution to the early current [ see also Kryshtal', Magura, and Parkhomenko (1969) and Chapter 5 of Khodorov (1974)] .  The data in Fig. 3-10f are from a plant cell,the fresh water alga <u>Nitella</u>.  This plant, which produces giant internodal cells with about the same dimension as the squid giant axon, has been described in detail by B. I. H. Scott (1962).  For <u>Nitella</u> it appears that the early current is carried by an outward flux of chloride ions, whereas the later current is primarily outward potassium.  The time required to relax from the $J_p$ branch to the $J_{ss}$ branch is the order of seconds for <u>Nitella</u> in contrast with a time of the order of milliseconds for the animal fibers in Figs. 3-10a-c.

No universally acceptable theory has yet been proposed to explain the relation between membrane electrodynamics (Fig. 3-10) and membrane biochemistry (Fig. 3-1).  An important recent contribution to this quest, however, is the review of various proposed mechanisms in Chapter 9 of the book by Khodorov (1974).  These mechanisms include: (a)  mobile carriers with affinities for particular ions, (b)  special pores with ionic selectivity and the ability to open and close, (c)  conformational changes in membrane macromolecules, and (d)  special mechanisms for artificial membranes.  Khodorov's discussion is particularly valuable because it considers the work of Russian scientists.

As Tasaki (1968, 1974b) and Changeux (1969) have shown, there is a considerable body of experimental evidence to suggest that the basic process of excitation in natural membranes involves a transition between two conformational states of the membrane. Fig. 3-10 certainly suggests the ubiquitous nature of two conductivity states, and more detailed data includes the following:

1. Direct observation of two conductivity states when $C_a^{++}$ is used as the external cation (Kobatake, and Tasaki,1973).

2. Observation of switching between these states by variation of the temperature.

3. Changes in extrinsic fluorescence during the time course of an action potential (Tasaki, 1974a).

4. Electron micrographs of configurational transitions involving collapse and extension of "headpiece stalks" in mitochondrial membranes (Hatase, et al. , 1972), and of lattice structure on electrically excitable membranes of insect photoreceptors (Gemme, 1969).

5. Nonaxoplasmic birefringence changes during the action potential (Cohen, Hille, and Keynes, 1970; Watanabe, Terakawa, and Nagano, 1973; Sato, Tasaki, Carbone, 1973).

6. Protein binding of a nontoxic dye during the action potential (Levine, Rozental', and Komissarchik, 1968).

7. Direct observation of spatial nonuniformity during switching of a squid axon (Inoue, Tasaki, and Kobatake, 1974).

Although Berestovskii, Liberman, Lunevskii, et al. (1970) report the absence of birefringe change in pure phospholipid bilayers, White (1975) finds evidence of a phase change at $17^\circ$C in gycerol monooleate bilayers from measurements plotting membrane capacitance against temperature. In any case, we can expect conformational states of protein complexes (Green, Ji, and Brucker, 1972) and of the proteins themselves (Frölich, 1970) to play a key role in membrane function.

It should be emphasized that the concept of a conformational change during activity of a natural membrane does not conflict with the idea that ions flow through channels or "pores" in the membrane; this point is discussed in detail by Hille (1970) (See also Ehrenstein, 1976). The two view points can be considered as complementary aspects of a more complex reality. On the other hand, one should not conclude that the switching observed on the leading edge in Fig. 1-3 is direct evidence of membrane macromolecular dynamics (Nachmansohn, 1959; Changeux, Thiery, Tungs, et al., 1967; Lehninger, 1968; Nachmansohm and Neumann, 1974). The basic positive feedback mechanism that drives an action potential is that discussed in the preceding paragraphs and diagrammed in Fig. 3-11 (Hodgkin, 1951, 1964). Several scientists have indicated how one might proceed from an essentially conformational membrane model to the ionic current data of Fig. 3-9, which, in turn, implies the feedback mechanism of Fig. 3-11 (Goldman, 1964, Jain, Marks; and Cordes, 1970; Chizmadzhev, Markin, and Muler, 1972, 1973).

In 1952 Hodgkin and Huxley introduced a phenomenological expression for the ion-current density through a squid membrane with the form

$$J_i = \overline{G}_K n^4 (v_{12} - V_K) + \overline{G}_{Na} m^3 h (v_{12} - V_{Na}) + G_L (v_{12} - V_L) \qquad (3.2.3)$$

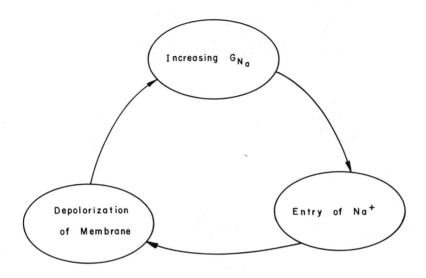

FIGURE 3.11.  Basic positive feedback mechanism for switching
of a squid nerve membrane (Hodgkin, 1964).

where $\overline{G}_K$ and $\overline{G}_{Na}$ are the maximum potassium and sodium conductances per unit area, respectively, and $G_L$ is a constant leak-
age conductance.  The phenomenological variables $n$, $m$, and $h$
lie between zero and unity; the potassium conductance is "turned
on" by $n$ and the sodium conductance is "turned on" and "turned
off" by $m$ and $h$, respectively.  It is assumed that $n$, $m$, and
$h$ are independently relaxing toward equilibrium values $n_o$, $m_o$,
and $h_o$ with characteristic times $\tau_n$, $\tau_m$, and $\tau_h$.  Thus

$$\frac{dn}{dt} = -\frac{n-n_o}{\tau_n} \; ; \qquad \frac{dm}{dt} = -\frac{m-m_o}{\tau_m}; \qquad \frac{dh}{dt} = -\frac{h-h_o}{\tau_h} \qquad (3.2.4a,b,c)$$

The relaxation parameters $(n_o, m_o, h_o, \tau_n, \tau_m,$ and $\tau_h)$ can be determined as functions of voltage such that (3.2.3) will reproduce
voltage-clamp data as in Fig. 3-9.  The nature of this functional
dependence is shown in Fig. 3-12a, where the constant values

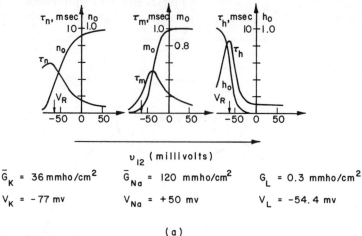

$\bar{G}_K = 36 \text{ mmho/cm}^2$   $\bar{G}_{Na} = 120 \text{ mmho/cm}^2$   $G_L = 0.3 \text{ mmho/cm}^2$

$V_K = -77 \text{ mv}$   $V_{Na} = +50 \text{ mv}$   $V_L = -54.4 \text{ mv}$

(a)

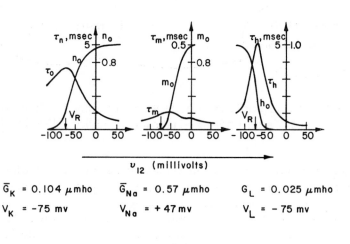

$\bar{G}_K = 0.104 \text{ } \mu\text{mho}$   $\bar{G}_{Na} = 0.57 \text{ } \mu\text{mho}$   $G_L = 0.025 \text{ } \mu\text{mho}$

$V_K = -75 \text{ mv}$   $V_{Na} = +47 \text{ mv}$   $V_L = -75 \text{ mv}$

(b)

FIGURE 3-12.   The functional dependence of Hodgkin-Huxley phenomenological parameters on membrane voltage: (a) for a typical squid axon; (b) for frog node of area $\sim 20 \text{ } \mu^2$ (redrawn from Cole (1968), pp. 283 and 479). Note: This data is for a temperature of 6.3°C. At other temperatures the time constants ($\tau$'s) should be divided by the factor $\kappa$ given in (3.2.6). Published in 1968 by the Regents of The University of California; reprinted by permission of the University of California Press.

given by Cole (1968) are also indicated.

When the corresponding variables are determined for the active node of a frog myelinated axon (area $\sim 20\ \mu^2$), the data are strikingly similar as shown in Fig. 3-12b. This result might be anticipated from a comparison of Figs. 3-10a and 3-10b.

Hodgkin and Huxley determined analytic expressions for the parameters in (3.2.3) of the form

$$\frac{dn}{dt} = \alpha_n (1-n) - \beta_n n$$

$$\frac{dm}{dt} = \alpha_m (1-m) - \beta_m m \qquad\qquad (3.2.4'a,b,c)$$

$$\frac{dh}{dt} = \alpha_h (1-h) - \beta_h h$$

Then, as functions of the voltage $v = v_{12} - V_R$ defined in (3.2.2) and measured in millivolts,

$$\alpha_n = 0.01\ \frac{10-v}{\exp(\frac{10-v}{10} - 1)}$$

$$\beta_n = 0.125\ \exp(\frac{-v}{80})$$

$$\alpha_m = 0.1\ \frac{25-v}{\exp(\frac{25-v}{10} - 1)}$$

$$\qquad\qquad (3.2.5a-f)$$

$$\beta_m = 4\ \exp(\frac{-v}{18})$$

$$\alpha_h = 0.07\ \exp(\frac{-v}{20})$$

$$\beta_h = \frac{1}{\exp(\frac{30-v}{10} + 1)}$$

where the units are $msec^{-1}$. Equations (3.2.5) give the rate constants measured at a temperature of 6.3°C. For other temperatures they should be multiplied by the factor $\kappa$ where

$$\kappa = 3^{(T-6.3)/10}$$

(3.2.6)

Clearly, (3.2.3) and (3.2.4) provide wide flexibility for fitting voltage-clamp data similar to that displayed in Figs. 3-8 and 3-9. Although the ion-battery potentials ($V_K$, $V_{Na}$, and $V_L$) are fixed by the respective concentration ratios, the maximum conductivities ($\overline{G}_K$, $\overline{G}_{Na}$, and $G_L$) can be adjusted in addition to the six functions of $v$ required to specify (3.2.4). Furthermore, the choice of powers appearing in (3.2.3) is somewhat arbitrary. The fourth power of $n$ was chosen to yield the "sigmoidicity" in the initial rise of potassium conductance evident from Fig. 3-8c and, as Hodgkin and Huxley note, "better agreement might have been obtained with a fifth or sixth power, but the improvement was not considered to be worth the additional complication." A later study by Cole and Moore (1960) suggested that the twenty-fifth power of $n$ is more appropriate in order to reproduce the time delay that appears when the membrane is switched on from the hyperpolarized state.[*] Similar considerations apply to the $m^3h$ factor in (3.2.3). The task is to represent a sodium conductance that first rises then falls (see Fig. 3-8b). Such an experimental result can be described by dependence on a single variable that obeys a second-order differential equation or on two variables, each obeying a first-order differential equation. Hodgkin and Huxley note, "the second alternative was chosen since it was simpler to apply the experimental results."

The Hodgkin-Huxley expression for ion current density (3.2.3) is well defined and useful for a variety of numerical and intuitive checks on experimental results. It has stimulated ever-widening analytical studies extending far beyond the professional scope of neurophysiology. Thus there is an inevitable (and regrettable) tendency to consider equations (3.2.3-5) as "graven on a stone tablet." The applied mathematician should be more concerned with the qualitative features of (3.2.3) than with the algebraic details. The biochemist, on the other hand, must concentrate on the development of a fundamental theory of membrane dynamics that can reproduce both the space- and voltage-clamped data. Useful reviews of the Hodgkin-Huxley equations are found in articles by Noble (1966) and Moore (1968) and books by Cole (1968)

---

[*] A suggestion that "wasn't recognized for tongue-in-cheek" (Cole, 1975).

and Khodorov (1974).

Various other suggestions for analytical representations of the potassium conductance include the work of:

1.  Tille (1965), who takes

$$I_K = \bar{G}_K N \qquad (3.2.7)$$

where

$$\frac{dN}{dt} = a_1 N + a_2 N^2 + a_3 N^3 + a_4 N^4 + a_5 N^5 \qquad (3.2.8)$$

and the $a_i$ are appropriately chosen functions of membrane voltage.

2.  FitzHugh (1965), who obtains (3.2.7) with $N = \exp(-\mu)$ and

$$\frac{d\mu}{dt} = \alpha - \beta\mu \qquad (3.2.9)$$

where $\alpha$ and $\beta$ are functions of the membrane voltage.

3.  Hoyt (1963), who uses (3.2.4a) from Hodgkin and Huxley (1952) then empirically determines the functional form for $G_K(n)$. She finds deviation from a power law at larger values of $n$.

4.  Rall and Shepherd (1968) and Goldstein and Rall (1974), who write

$$J_i = v - E(1 - v) + I(v + 0.1)$$

$$\frac{dE}{dt} = k_1 v^2 + k_2 v^4 - k_3 E - k_4 I \qquad (3.2.10a,b,c)$$

$$\frac{dI}{dt} = k_5 E + k_6 EI - k_7 I$$

Because $k_1, \ldots, k_7$ are constant, numerical integration is simpler than with the Hodgkin-Huxley equations. Goldstein and Rall (1974) suggest several sets of values for these constants to represent nerves of the squid, lobster, and crab.

In a discussion following the presentation of FitzHugh (1965), Cole points out the wide range of functional expressions that can represent the sigmoid nature of the potassium conductance rise with roughly equal accuracy.  During this discussion Cole, Hoyt, and FitzHugh are in agreement that there is no uniquely superior analytical form.  For the Purkinje fibers in the mammalian heart, however, Noble (1962) has described a modified representation for the potassium conductance that account for the slow recovery indicated in Fig. 3-10d.  This slow recovery is necessary for the generation of heartbeats.

Analytical study of the rise and fall of sodium conductance (Fig. 3-8b) has been of more fundamental importance.  Frankenhaeuser and Huxley (1964) have shown for myelinated axons of the toad (Xenopus laevis) that an $m^2h$ dependence is more appropriate.  Hoyt (1963, 1968) and Hoyt and Adelman (1970) have demonstrated that for a squid giant axon the sodium conductance is somewhat better represented by dependence on a single variable that satisfies a second-order differential equation or, equivalently, two variables that satisfy coupled first-order equations.  Hoyt and Adelman state, "These conclusions imply that the mechanism responsible for the increase in sodium conductance is more likely to be dependent upon the production of an intermediate state than on the competition of two antagonistic but independent processes. . ." However, see also Jakobsson (1973).  Molecular theories leading to coupled equations include the works of Mullins (1959), Goldman (1964), Fishman, Khodorov, and Volkenshtein (1972), and Chizmadzhev, Markin, and Muler (1972, 1973).  Other models for membrane dynamics with varying degrees of phenomenology and membrane biochemistry include the work of Jain, Marks, and Cordes (1970), Offner (1970, 1972, 1974), Moore and Jakobsson (1971), Jakobsson and Scudiero (1975) and Mackey (1975).  But it is important to emphasize that all the space clamped data should be represented by a viable biochemical theory.  The kinetic model of Jain, Marks, and Cordes (1970), for example, provides a parsimonious description of the data in Figs. 3-9 and 3-10a by assuming discrete membrane sites that can make transitions first to states of sodium conductivity and then to potassium.  Only six rate constants are assumed, four of which are constant, and the remaining two are related in a natural way to the notion of a confirmational change in the membrane.  But since the input and output to the sodium conducting states are first-order transitions with constant (i. e. , voltage independent) rates of 4500 $sec^{-1}$ and 2000 $sec^{-1}$, respectively, the time derivative of sodium conductivity is constrained by the model to remain constant over times of the order of 0.1 msec. The measurements of Hodgkin and Huxley (1952b, Fig. 5), however, clearly demonstrate a jump of $(dG_{Na}/dt)$ from positive to negative values in a time that is short compared with 0.1 msec when a stimulating voltage pulse is suddenly reduced to zero. * Whether the model can be modified to eliminate this objection is not yet clear.

For a space-clamped squid axon membrane between 15°C and 38°C, Cole, Guttman, and Bezanilla (1970) have demonstrated that the relationship between stimulus (S) and maximum response (R) is a single valued, monotone increasing curve as is indicated in Fig. 3-13. Defining an amplification factor

$$A \equiv \frac{dR}{dS}\bigg|_{max} \tag{3. 2. 11}$$

it is clear that A rises dramatically with decreasing temperature. Calculations of A for various temperatures can easily be made from the Hodgkin-Huxley equations, together with the adjustment of rates indicated in (3. 2. 6). Some results of these calculations are as follows (Cole, Guttman, and Bezanilla, 1970):

| T(°C) | 6. 3 | 15 | 25 | 35 | 45 |
|---|---|---|---|---|---|
| A | $6 \times 10^{15}$ | $1.6 \times 10^{8}$ | $2.2 \times 10^{3}$ | 21 | 3.1 |

The value of A predicted at 6. 3°C is far too large to be checked experimentally; noise effects would mask any attempt to determine whether the R(S) curve is really single valued. Above 15°C, however, the Hodgkin-Huxley predictions are confirmed by experiment, and there is no evidence to contradict the 6. 3°C value of A. This result has led Cole, Guttman, and Bezanilla (1970) to conclude that the "all-or-none" law does not apply to a space

---

* The present author is indebted to A. F. Huxley for pointing out this difficulty with the model of Jain, Marks, and Cordes (1970).

clamped patch of normal membrane and to offer the following advice to membrane theorists:

1. Do not attack the problem of all-or-none propagation of a nerve inpulse directly; it is unnecessarily difficult and all of the important facts are explained by the ion conductances.

2. Do not invoke mechanisms which will produce an all-or-none response in the far simpler situation of membrane uniformity; the conclusion is contrary to the best available evidence.

3. Address the problem of explaining the Hodgkin-Huxley conductances; these are experimental facts of normal axon membranes on which most of the important and interesting behaviors of excitable membranes and axons depend.

Thus the challenge remains to determine what is really happening during "sodium turn-off. " What physical process does "Nat" (Fig. 3-8) represent? An interesting suggestion, which has been offered independently by Hoyt and Strieb (1971) and by Landowne (1972), is that sodium current is carried almost entirely by those ions initially stored within the membrane. Sodium turn-off then arises naturally as a depletion of the stored ions that can be replenished only from the external solution on a longer time scale. This hypothesis implies that the temperature dependence of membrane permeability to sodium ions should be weak, whereas (3. 2. 6) indicates a rather strong temperature dependence. Assuming the rates inversely proportional to permeability, (3. 2. 6) would imply a threefold permeability increase for a $10^{0}$ decrease in temperature. Experiments by Landowne (1973) and Cohen and Landowne (1974) using $^{22}Na$ as a radioactive tracer indicate, however, a temperature dependence of sodium permeability that is much weaker.

Cohen and Landowne emphasize that these results do not invalidate the predictive power of the Hodgkin-Huxley equations, but merely imply that the temperature dependence of the rates cannot be ascribed to permeability changes. Thus what else could cause the rates to change? Landowne (1972) has suggested that sodium and potassium ions may pass through the membrane under the restrictions of "space charge limited flow. " Although originally proposed to describe electron currents in a vacuum diode

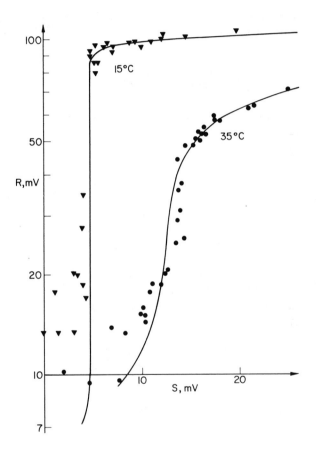

FIGURE 3-13. Comparisons of maximum response (R) with amplitude of stimuli (S) for space-clamped squid membranes at 15°C and 35°C. The curves are calculated from the Hodgkin-Huxley equations (Cole et
al. , 1970).

(Compton and Langmuir, 1931; M. I. T. , 1943), space-charge limited
flow is also observed in the conductors of solid-state electronics
(van der Ziel, 1957).   For current flow between two electrodes in
which the velocity of the charge carriers depends on the voltage
as

$$\text{velocity} = kv^a \tag{3.2.12}$$

Poisson's equation and the assumption of a "virtual cathode" near
one electrode leads to a generalized form of Child's law

$$J = \frac{2(a-1)}{(a+1)^2} \frac{k\epsilon}{\delta^2} v^{a+1} \tag{3.2.13}$$

where $\delta$ is the electrode separation.   Thus Landowne writes the
ion currents in the form

$$J_i = K v^{a+1} \tag{3.2.14}$$

where K is a constant.   The transit time, $\tau$, is inversely proportional to carrier velocity and is thus approximated as

$$\tau = A v^{-a} \tag{3.2.15}$$

where A is another constant.   From (3.2.14) the ionic conductances should be given by

$$G_i = K v^a \tag{3.2.16}$$

Landowne has used (3.2.15) and (3.2.16) to normalize the Hodgkin-
Huxley conductance data in Fig.  3-8.   He writes

$$G_i' = \frac{G_i(t) - G_i(0)}{K v^a} \tag{3.2.17}$$

$$t' = \frac{t}{A v^{-a}} \tag{3.2.18}$$

with the following choice of constants:

|       | Potassium | Sodium |                   |
|-------|-----------|--------|-------------------|
| K =   | 1. 3      | 0. 21  | $mmho/cm^2$       |
| A =   | 49        | 49     | $(mV)^a$ sec      |
| a =   | 0. 6      | 1. 0   | - - -             |

Then the curves A, B, D, and F of Fig. 3-8c appear as in Fig. 3-14a, while the curves of B, C, D, F, and G from Fig. 3-8b appear as in Fig. 3-14b.

Whether this normalization is mere happenstance or, as Landowne suggests, a clue to the dominating physical mechanism remains to be determined. The choice of a = 1. 0 for sodium ions may appear somewhat disturbing since (3. 2. 13) then implies that the constant K should equal zero. But K is found to be considerably smaller for the sodium ions than for potassium, and a could be changed by a few percent in (3. 2. 17) and (3. 2. 18) without introducing a great change into Fig. 3-14.

Strandberg (1976) has taken a more direct physical approach to the construction of a phenomenological theory. For both the sodium and the potassium ions he assumes two ensembles of states (conducting and nonconducting), each distributed over energy in a quasicontinuous manner. Such a distribution of energy states does not seem unreasonable because the large protein molecules that facilitate ionic conduction have many degrees of freedom. Strandberg shows under rather general assumptions that potassium conductance should be expected to vary with applied voltage as

$$G_K = \frac{\overline{G}_K}{1 + \exp\{-[\Delta E + \Delta F(v)]/kT\}} \qquad (3. 2. 19)$$

where $\Delta E$ is the unperturbed mean energy difference between conducting and nonconducting conformational states and $\Delta F(v)$ is the contribution to this difference from the external voltage. With

$$\overline{G}_K = 22. 5 \ mmho/cm^2$$

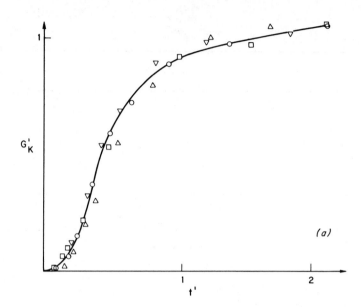

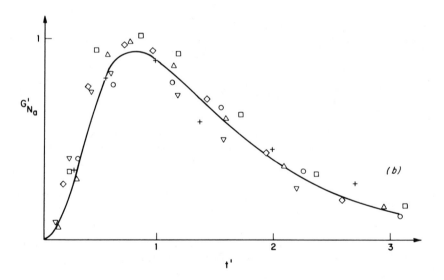

FIGURE 3-14.  Normalization of the Hodgkin-Huxley conductance
data in Fig. 3-9 by Landowne (1972);
(a) potassium conductance: O - A, △- B, ▢ - D, ▽- F;
(b) sodium conductance: O - B, ▢ - C, ◇- D, △- F,
▽ - G.

$$\Delta E = 23 \text{ meV}$$

$$\Delta F(v) = [0.8 v_{12} - 9.4 \exp (-v_{12}/11.9)] \quad \text{meV}$$

a rather good fit to the Hodgkin-Huxley steady-state potassium conductance [ see (3.2.3)]

$$G_K = \bar{G}_K n_o^4 (v) \qquad (3.2.20)$$

is obtained. The point of Strandberg's work, however, is not simply to introduce a new parametric representation of old data. Dynamic effects can be treated as relaxations between the steady-state distributions that obtain at different voltages, and the temperature dependence of the rates may suggest acceptable theories for the nature of the conduction process. Furthermore, (3.2.19) implies a temperature dependence of steady-state conductance that varies with voltage in a rather complex way. In particular, as $G_K \rightarrow \bar{G}_K$, it becomes insensitive to temperature, which is in rough agreement with the observations of Landowne (1973) and by Cohen and Landowne (1974). Predictions of: (a) entropy generation and heat flow, (b) pressure dependence of membrane free energy, (c) effects of dissolved inert gas, (d) extra low frequency (ELF) electrical polarization, and (e) membrane impedance, all would seem to be feasible from this thermodynamic approach. Many new opportunities thus arise for the classical physicist to contribute to the solution of the membrane riddle.

# 4

# The Nerve Fiber

···· our senses have widened. Membranes, webs
of nerves that lay white and limp, have filled and
spread themselves and float round us like filaments,
making the air tangible and catching in them far-away
sounds unheard before.

Virginia Woolf


We are now ready to begin the discussion of level 3, namely,
the development of neuron theory from membrane dynamics and
electromagnetic theory. Our first problem is to understand the
propagation of a solitary wave or action potential (see Fig. 1-3)
along one of Galvani's oily tubes. These solitary waves are stable spatiotemporal entities that arise as solutions to the diffusion
equation (2.30) when it is rendered nonlinear through a suitable
representation of the ion current $j_i$.

In retrospect, it seems that applied mathematicians forewent
an unusual opportunity to make important scientific contributions
by ignoring the study of the nonlinear diffusion equation. One ex-
ception to this generalization was the work by Kolmogoroff,
Petrovsky, and Piscounoff (1937) with the equation[*]

$$\phi_{xx} - \phi_t = F(\phi) \qquad (4.1)$$

related to the biological problem of genetic diffusion. These

---

[*] Equation (4.1) should perhaps be called the K. P. P. equation.

authors showed how steplike initial conditions would evolve into a solitary wave solution of the form

$$\phi(x, t) = \phi_T(x - ut), \quad u = \text{const} \tag{4.2}$$

developed phase-plane techniques for determining $\phi_T$, and derived explicit formulas for the traveling-wave velocity, u. This important contribution was completely overlooked by electrophysiologists in the United States; indeed, it is not even noted in the otherwise exhaustive bibliography of the book by Cole (1968). The failure of applied mathematicians to undertake a timely study of (4.1) cannot be ascribed to technical inefficiency in the face of the "enormous mathematical difficulties" envisaged by Hermann (1905). The studies by Boussinesq (1872) and by Korteweg and deVries (1895) of the hydrodynamic solitary waves described by Scott Russell (1844) indicate that there was ample understanding of nonlinear PDE's even before the turn of the century. As Cohen (1971) has suggested, the difficulty may have been the assumption by most mathematicians that the diffusive and nonpropagating behavior of linear diffusion equations would carry over to the nonlinear case.

Yet one need not turn to Hermann's line of burning powder or the Japanese incense mentioned by Kato (1924) for a clear physical representation of nonlinear diffusion; the ordinary candle had been lighting scientific study tables for centuries. Diffusion of heat down the candle releases wax to the flame where it burns to supply the heat. If P is the power (J/sec) necessary to support the flame and E is the chemical energy stored per unit length of the candle (J/m), then the flame (nonlinear wave) will travel at the velocity u for which

$$P = uE \tag{4.3}$$

The rate at which energy is "eaten" (uE) must equal the rate at which it is "digested" by the flame (P). Equation (4.3) is of more than pedagogical interest; when we turn to the development of formulas for the calculation of nerve-pulse propagation velocity, we use (4.3) to find solutions of (4.1) with the traveling-wave character indicated in (4.2).

Nonlinear wave problems can be divided into two main classes: a) open systems for which solitary traveling waves imply a balance between rate of energy release by the nonlinearity and its consumption as is indicated by (4. 3) and (b) closed systems for which energy is conserved through a conservation law

$$\mathcal{E}_t + \mathcal{P}_x = 0 \tag{4.4}$$

where $\mathcal{E}$ is energy density and $\mathcal{P}$ is the power flow. Wave problems of class b) include the hydrodynamic waves that were studied by Boussinesq (1872) and by Korteweg and deVries (1895). In this case solitary waves involve a balance between the effects of nonlinearity and dispersion, and the propagation velocity is an adjustable parameter in a family of solutions. Such energy-conserving solitary waves sometimes exhibit and infinite number of conservation laws and the nondestructive collisions characteristics of "solitons." Nothing further will be said here about class b); the interested reader is referred to the paper by Scott, Chu, and McLaughlin (1973) for a review of this research. Although the present discussion concentrates on nonlinear wave problems of class a), it should not be assumed that conservation laws are unimportant. Indeed, we find in Chapter 5 that an approximate conservation law for electric charge can be useful in determining the conditions necessary to stimulate a nerve fiber to the threshold of excitation, and also in Chapter 6 that a conservation law for pulses may help to analyze the evolution of a pulse burst along a fiber.

## 1.  THE HODGKIN-HUXLEY AXON

Let us consider the nonlinear dynamics of the nerve fiber shown in Fig. 2-2a. The first-order partial differential equations are (2. 21), together with (3. 2. 4). Combining (2. 21b) with (2. 29) we can write these as

$$
\begin{array}{c}
\dfrac{\partial v}{\partial x} = -r_s i \\[2mm]
\dfrac{\partial i}{\partial x} + c\dfrac{\partial v}{\partial t} = -j_i(v, n, m, h) \\[2mm]
\dfrac{\partial n}{\partial t} = -\dfrac{n - n_o(v)}{\tau_n(v)}
\end{array}
\tag{4.1.1a-c}
$$

$$\frac{\partial m}{\partial t} = - \frac{m - m_o(v)}{\tau_m(v)}$$

$$\frac{\partial h}{\partial t} = - \frac{h - h_o(v)}{\tau_h(v)}$$

(4.1.1d, e)

where $j_i$ in (4.1.1b) is the membrane ion current per unit length. From here on it is typographically convenient to use the voltage variable $v = v_{12} - V_R$ defined in (3.2.2); evidently this makes no difference on the left-hand sides of (4.1.1a) and (4.1.1b). From (3.2.3)

$$j_i = \bar{g}_K n^4 (v + V_R - V_K) + \bar{g}_{Na} m^3 h(v + V_R - V_{Na}) + g_L(v + V_R - V_L)$$

(4.1.2)

where $\bar{g}_K = 2\pi a \bar{G}_K$, $\bar{g}_{Na} = 2\pi \bar{G}_{Na}$, and $g_L = 2\pi a G_L$.

The "average axon" chosen for numerical study by Hodgkin and Huxley (1952) had the following parameters in addition to those specified in the previous section:

| Resting Potential | Axoplasm Conductivity | Axon Radius | Membrane Capacitance |
|---|---|---|---|
| $V_R = -65\,mV$ | $\sigma = 2.9\,mho/m$ | $a = .238mm$ | $C = 1\,\mu fd/cm^2$ |

One approach to the analysis of these equations is to seek traveling-wave solutions where all dependent variables (v, i, h, m, and h) are functions only of a moving spatial variable

$$\xi = x - ut$$

(4.1.3)

This can be considered as a special case of the more general independent variable transformation

$$x \rightarrow \xi = x - ut \qquad \frac{\partial}{\partial x} \rightarrow \frac{\partial}{\partial \xi}$$

$$\text{so} \tag{4.1.4}$$

$$t \rightarrow \tau = t \qquad \frac{\partial}{\partial t} \rightarrow \frac{\partial}{\partial \tau} - u \frac{\partial}{\partial \xi}$$

Assuming independence with respect to $\tau$ in the $(\xi, \tau)$ system we can replace $(\partial/\partial x)$ by $(d/d\xi)$ and $(\partial/\partial t)$ by $(-u\, d/d\xi)$, whereupon (4.1.1) become the ordinary differential equations

$$\frac{dv}{d\xi} = - r_s i$$

$$\frac{di}{d\xi} = - r_s cui - j_i$$

$$\frac{dn}{d\xi} = \frac{n - n_o}{u\tau_n} \tag{4.1.5}$$

$$\frac{dm}{d\xi} = \frac{m - m_o}{u\tau_m}$$

$$\frac{dh}{d\xi} = \frac{h - h_o}{u\tau_h}$$

This is an _autonomous_ set of equations (Hurewicz, 1958; Lefschetz, 1962) since the derivatives are uniquely defined as functions of the dependent variables. Thus phase-space techniques can be helpful in understanding the structure of solutions (Kolmogoroff, Petrovsky, and Piscoundoff, 1937). It is important to note, however, that u [the velocity of the moving spatial coordinate in (4.1.4)] appears as an adjustable parameter in (4.1.5). In general, one can expect the topological character of the phase-space trajectories to depend on the value chosen for the velocity u. Only those trajectories for which the dependent variables are bounded are of physical interest here. In particular, a trajectory corresponding to the action potential shown in Fig. 1-3 should have the qualitative character indicated in Fig. 4-1. The values $v = 0$, $i = 0$, and $(n, m, h) \doteq (.35, .06, .6)$ are a solution of (4.1.1) so the corresponding point in the phase space of (4.1.5) is a _singular_ point at which all the $\xi$-derivatives are equal to zero. The task of finding a pulse-like traveling-wave solution for (4.1.1) involves determining the proper value of the velocity u at which a trajectory emanating from this singular point (at $\xi = -\infty$)

eventually returns to it (as $\xi \to +\infty$).  Such a trajectory is sometimes called <u>homoclinic</u>, while a <u>heteroclinic</u> trajectory passes
between two different singular points.

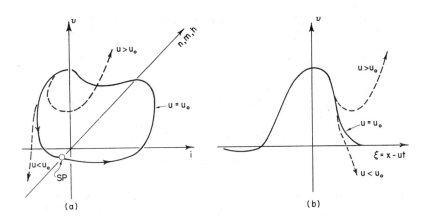

FIGURE 4-1.    (a)  Phase-space trajectory corresponding to (b) an
               action potential.  The phase space has five dimen-
               sions, but  n , m, and  h  are indicated along a
               single axis.

A homoclinic trajectory was determined by Hodgkin and Huxley
(using a hand calculator) in 1952.  Voltage and membrane conductance are plotted as a function of time from this calculation in Fig.
4-2 to obtain the proper value of  u  equal to 18. 8 m/ sec.  This
is in satisfactory agreement with the measured value of 21.2 m/ sec;
and, as shown by a comparison of Figs. 1-3  and 4-2, so also are
the waveforms  v(t)  and  G(t).
    Theoretically, the discovery of a pulse-like traveling-wave
solution for (4. 1. 1) from an investigation of the phase-space to-
pology associated with (4. 1. 5) does not imply that the pulse is
stable to perturbations of its shape.  Such <u>waveform instability</u> involves dependence on  $\tau$ , and (4. 1. 5) was derived with the specific assumption of independence with respect to  $\tau$ .  (We study this
question in detail in Chapter 5. ) Another form of instability that

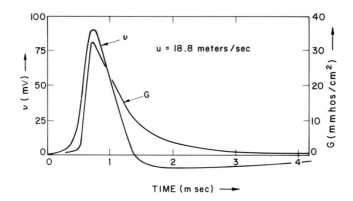

FIGURE 4-2.   Waveforms of the action potential and membrane
              conductance calculated from (4. 1. 5) at 18. 5°C
              (redrawn from Hodgkin and Huxley, 1952d).

appears in these calculations is <u>numerical instability</u> during the
integration of (4. 1. 5).   This arises because the assumed pulse
velocity, u , is an adjustable parameter in the analysis.   Choosing
u  slightly too small or too large may cause the computed waveform to diverge (see Fig. 4-1).   As we see later, such numerical
instability of a solution to  (4. 1. 5) seems to be a necessary condition to avoid a waveform instability in the corresponding solution
of (4. 1. 1).

     Machine computations for the space clamped membrane were
first reported by Cole, Antosiewicz, and Rabinowitz (1955), and
for the propagating axon, by FitzHugh and Antosiewicz (1959) and
Huxley (1959).   Huxley demonstrated the existence of a second
pulse solution (shown in Fig. 4-3) that propagates with only 30%
of the velocity of the full action potential.   This pulse has an unstable waveform; it will either decay to zero or rise to the full action potential and thus represents a boundary or threshold state of
the fiber.   Huxley (1959) also indicated the possibility of a periodic

wave train that would correspond to a closed cycle in the phase
space scetched in Fig. 4-1.   The observation of a threshold pulse
was confirmed by Cooley and Dodge (1966) through direct integration of (4.1.1), who extended the result by assuming that the effect of a narcotic agent would be to lower $\bar{g}_{Na}$ and $\bar{g}_K$ by a factor $\eta$.   The results are plotted in Fig. 4-4, where it can be seen
that no attenuationless propagation or threshold effect obtains for
$\eta < \eta_c = .261$.   At smaller values of this "narcotization factor" a
"decremental" pulse (Lorente de No and Condouris, 1959) propagates with slowly diminishing amplitude as shown in Fig. 4-28.
Since this pulse is not a function only of the argument $x - ut$, it
is not represented by solutions of (4.1.5) and requires the complete set (4.1.1) for its description.

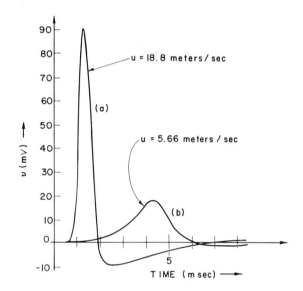

FIGURE 4-3.   (a)  A full-sized action potential and (b)  an unstable
threshold pulse for the Hodgkin–Huxley axon at
18.5°C   (redrawn  from Huxley, 1959).

Impedance bridge measurements by Cole and Baker (1941) indicated that the membrane appears to have an inductive current
component at small  ac  amplitudes between 30 Hz  and 200 kHz.

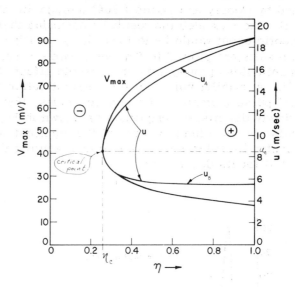

FIGURE 4-4.    Amplitude and velocity for a traveling wave pulse on
a Hodgkin-Huxley axon plotted against a "narcotization factor, " $\eta$, which reduces the sodium and potassium conductances (redrawn from Cooley and
Dodge, 1966).

For the membrane equivalent circuit representing $1 \text{ cm}^2$ of membrane shown in Fig. 4-5a, they found $C = 1 \mu F$, $R = 400$ ohm,
and $L = 0.2$ H. Hodgkin and Huxley (1952) investigated the dynamical relation between small changes in voltage and current in
(3.2.3) and directly calculated the values $R = 820 \Omega$ and $L = 0.39$
H with a threefold increase in $L$ for a $10^\circ$ fall in temperature.
Such an inductance is much too large to have any connection with
magnetic fields, and a physical interpretation is illustrated in
Fig. 4-5b that is contingent on the experimental fact that membrane conductance [$G$ in (3.1.12)] remains constant for times of
the order of $\leq 100 \mu\text{sec}$ (Mauro, 1961). If the current curve is
concave in the direction of depolarization, a sudden change of
current from $J_1$ to $J_2$ must be associated with a change of voltage from $v_1$ to $v_2'$. The voltage will then slowly relax toward a

smaller difference $v_2$.  These conditions are met by the  n  and  h  dependencies in (3. 2. 3), both of which contribute to the inductance indicated in Fig. 4-5a.  Extensive studies of this "phenomenological" inductance include those by Chandler, FitzHugh, and Cole (1962) and Mauro, Conti, Dodge, et al. (1970).  Recently Guttman, Feldman, and Lecar (1974) have measured squid membrane response to various levels of white noise from which they have computed the cross correlation of input with response.  Again a parallel RLC representation of the membrane seemed appropriate (Fig. 4-5a) with a resonant frequency varying from about 100 Hz at 10°C to 250 Hz  at 20°C.  This approach has the experimental advantage that response is measured simultaneously at all frequencies, thus eliminating errors caused by axon fatigue.

The phenomenological inductance also influences the propagation of alternating subthreshold waves on the axon; this is evident from the "overshoot" in the return to rest of the action potential in Fig. 4-2.  Subthreshold oscillatory propagation has been studied in detail by Sabah and Leibovic (1969), and Leibovic and Sabah (1969), and Leibovic (1972) using Laplace transform techniques and by Mauro, Freeman, Cooley, et al. (1972), who use both numerical analysis of (4. 1. 1) and experimental observations on squid axons to show that phase velocity of an oscillatory subthreshold wave is rather closely related to the pulse velocity of an potential.  Opatowski (1950) has also studied this relation.  In electronic jargon the squid axon resembles a low  Q, bandpass filter tuned to about 100 Hz when it is stimulated by a subthreshold, oscillatory current.

Cooley and Dodge (1966) also computed the response of a Hodgkin-Huxley axon to a steady stimultation by longitudinal current [ i(0, t) = const  in Fig. 2-2a] .  For a steady current around 3. 4 μA, a periodic train of spikes was generated with a frequency rather insensitive to the stimulation.  This result is in contrast to the real axon, which generates a burst of spikes.  FitzHugh (1969) has suggested that the real axon exhibits an "adaptation" effect that tends to decrease excitability with a time constant of the order of a second.  Such an effect, which is not represented by the Hodgkin-Huxley equations, may be connected with slow changes in ion concentration or in temperature.

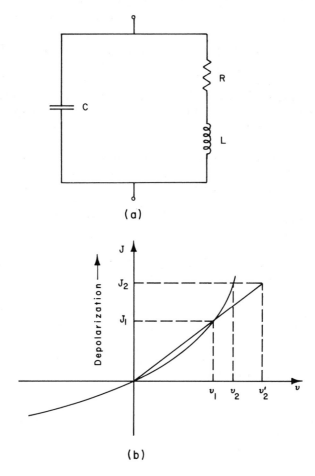

(a)

(b)

FIGURE 4-5.    (a)  Membrane small signal equivalent circuit
measured by Cole and Baker (1941); (b) physical
explanation for the phenomenological inductance.


## 2.    PROPAGATION OF THE LEADING EDGE

Comparison of numerical results reported in the previous section with corresponding experimental data indicates that the
Hodgkin-Huxley equations (4. 1. 1) are of considerable value in describing the facts of electrophysiology, but it is also of interest

to consider approximate forms that are amenable to analytic investigation. Physical motivation for one such approximation stems from the following observations (see Fig. 4-2):

1.  The most rapid dynamical change occurs on the leading edge of an action potential.

2.  This leading edge transition carries the membrane potential from its resting potential to approximately the sodium diffusion potential, $V_{Na}$.

3.  The velocity of the leading edge determines the velocity of the entire action potential.

For the squid giant axon the functions $n_o, m_o, h_o, \tau_n, \tau_m$, and $\tau_h$ are sketched in Fig. 3-12a from which it is evident that the relaxation time, $\tau_m$, for sodium turn-on is about an order of magnitude less than $\tau_n$ and $\tau_h$ for potassium turn-on and sodium turn-off, respectively. Thus it is interesting to consider the approximation (FitzHugh, 1969)

$$\tau_m = 0, \qquad \tau_n = \tau_h = \infty$$

where the ion current through the membrane (4.1.2) becomes simply a function of voltage $j_i \approx j(v)$ with

$$j(v) = \bar{g}_K n_o^4 (V_R)(v + V_R - V_K) + \bar{g}_{Na} m_o^3(v) h_o (V_R)(v + V_R - V_{Na})$$

$$+ g_L(v + V_R - V_L) \qquad\qquad (4.2.1)$$

This approximation is valid only for dynamic processes that occur in times long compared with $\tau_m$ and short compared with $\tau_n$ and $\tau_h$, but, as reference to Fig. 4-2 indicates, the leading edge transition comes close to fulfilling these requirements. Equation (2.30) then takes the form[*]

$$v_{xx} - r_s c v_t = r_s j(v) \qquad\qquad (4.2.2)$$

---

[*] From this point on in the present text the conventional subscript notation for partial differentiation is used wherever it is typographically convenient.

which is the K. P. P. equation for nonlinear diffusion. Together
with (4. 2. 2) it is convenient to write (2. 21) in the form

$$v_x = -r_s i$$

$$i_x + c v_t = -j(v)$$

(4. 2. 3a, b)

as an equivalent set of first-order PDEs.

Equation (4. 2. 1) does not have a particularly convenient analytic form, but we expect it to go through zero at the origin (the
resting potential) at a higher voltage $V_2 = V_{Na} - V_R$, and at a
voltage, $V_1$, somewhere between. With this in mind, let us apply
the transformation (4. 1. 4) discussed in the previous section to
(4. 2. 3) with the assumption that $(\partial/\partial\tau) = 0$. Then the set of
ordinary equations that are equivalent to (4. 1. 5) becomes

$$\frac{dv}{d\xi} = -r_s i$$

$$\frac{di}{d\xi} = -r_s cui - j(v)$$

(4. 2. 4a, b)

Singular points in the $(v, i)$ phase plane for this set occur where
$i = 0$ and $j(v) = 0$, that is, at $v = 0$, $V_1$ and $V_2$. If we define

$$g(v) \equiv \frac{dj}{dv}$$

(4. 2. 5)

then $g(0)$ and $g(V_2)$ will be positive and $g(V_1)$ will be negative
(see Fig. 4-6a). From this one can show (Scott, 1962) that the
singular points at $(i, v) = (0, 0)$ and $(0, V_2)$ are saddle points,
while the intermediate singular point at $(0, V_1)$ is an inward (outward) node or focus for $u > 0$ $(< 0)$. Kunov (1967) used "Bendixon's
negative criterion" (Andronov, Vitt, and Khaikin, 1966) to show that
(4. 2. 4) has a homoclinic trajectory, corresponding to a "pulse-
like" solution of (4. 2. 3), only for zero velocity. Thus the basic
solutions with nonzero velocity are the "level-change" waves
shown in Fig. 4-6b. From the phase-space point of view, the
velocity of such a transition is fixed by the condition that an isolated trajectory leaving one saddle point (at $\xi = -\infty$) must become
an isolated trajectory approaching the other saddle point (as $\xi \to \infty$).

Yoshizawa (1971) has demonstrated that these waves can either charge the membrane capacitance when area $A_2$ is greater than area $A_1$, or discharge the capacitance for $A_1 > A_2$. In either case the power balance condition (4. 3) must be satisfied.

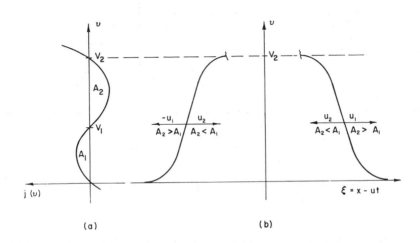

FIGURE 4-6.   (a)  A representation of $j(v)$ as in (4. 2. 1);
(b)  propagating waves that change the voltage level.

If $A_1 = A_2$, these velocities are equal to zero, which is a special case of the zero velocity pulse indicated in Fig. 4-7 for the case $A_2 > A_1$. From (4. 2. 4) with $u = 0$, it is easily seen that a pulse like solution is obtained by substituting into (4. 2. 4a) the homoclinic trajectory

$$ i = \pm \left[ \frac{2}{r_s} \int_0^V j(v') \, dv' \right]^{\frac{1}{2}} \tag{4. 2. 6} $$

Although this solution is unstable, as we see in the following paragraphs, it is of interest because it specifies the condition for threshold stimulation of a fiber. Lindgren and Buratti (1969) have shown the pulse velocity to be nonzero for a tapered fiber.

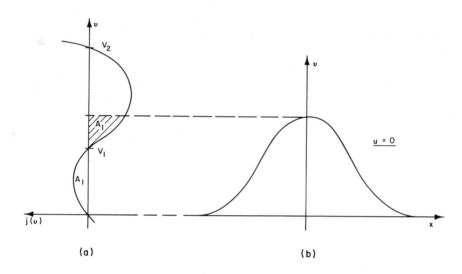

FIGURE 4-7.    (a)  j(v)  with  $A_2 > A_1$;   (b)   stationary pulse
solution.

A family of analytic solutions for the wave forms and velocities indicated in Fig.  4-6 can be obtained by writing (Scott, 1976)

$$\frac{dv}{d\xi} = T(v) \qquad (4.2.7)$$

where (4.2.4) requires that  T  must satisfy

$$T' = r_s \frac{j(v)}{T} - r_s cu \qquad (4.2.8)$$

For  u = 0,  the pulse-like trajectory of (4.2.6).   Now suppose
u ≠ 0   and

j(v)  is a polynomial of order  n,  and

T(v)  is a polynomial of order  m

Then  T'  is of order  (m-1)  and from (4.2.8)

$$n = 2m - 1 \qquad (4.2.9)$$

The case $m = 2$ implies $n = 3$ so $j(v)$ must be approximated by a cubic polynomial (Nagumo, Yoshizawa, and Arimoto, 1965; FitzHugh, 1969)

$$j(v) = Bv(v - V_1)(v - V_2) \qquad (4.2.10)$$

where $B$ is a constant (with units of $mho/V^2$) chosen to make $j(v)$ approximate $2\pi a J_p$ from Fig. 3-10a or (4.2.1). Since $m = 2$, a suitable quadratic trajectory is

$$i = Kv(v - v_2)$$

which on differentiation gives

$$\frac{di}{dv} = 2Kv - KV_2$$

However, $(di/dv)$ can also be evaluated by dividing left- and right-hand sides of (4.2.4) to obtain

$$\frac{di}{dv} = cu + \frac{B}{r_s K}(v - V_1)$$

Thus $K = -(B/2r_s)^{\frac{1}{2}}$ so

$$u = \sqrt{\frac{B}{2r_s c^2}}(V_2 - 2V_1) \qquad (4.2.11)$$

and (4.2.4a) can be integrated to the logistic function

$$v = \frac{V_2}{1 + \exp\left[(Br_s/2)^{\frac{1}{2}}(x - ut - x_o\right]} \qquad (4.2.12)$$

Note that the velocity given by (4.2.11) changes sign as $V_1$ becomes greater than $(V_2/2)$. This corresponds to the area condition indicated on Fig. 4-6b. Similar results have been obtained for other nonlinear wave systems simulating the nerve axon by Il'inova and Khokhlov (1963) and by Parmentier (1969).

Another approximation for $j(v)$ that permits an analytic solution for (4.2.4) corresponds to the case $m = 1$, so from (4.2.9) $n = 1$ and we have a piecewise linear curve indicated in Fig. 4-8. Below a voltage $V_1$ the membrane is assumed to remain in a _resting_ state with low conductance; above $V_1$ it is assumed to switch into an _active_ state of much higher conductance. Such an approximation is certainly suggested by several of the curves plotting $J_p$ against $v_{12}$ in Fig. 3-10. Using the notation of Tasaki (1968), we write

$$j(v) = g_r v \qquad \text{for} \quad v < V_1$$
$$= g_a(v - V_2) \quad \text{for} \quad v > V_1 \qquad (4.2.13)$$

The discontinuity at $V_1$ is acceptable because (4.2.2) and (4.2.3) do not involve derivatives of $j(v)$. With $j(v)$ approximated as in (4.2.13), (4.2.2) is linear both above and below $V_1$. Thus the nonlinearity in the problem manifests itself only where $v = V_1$. To simplify the discussion we begin by assuming that $g_r = 0$. Equation (4.2.4) can be written

$$\frac{d^2v}{d\xi^2} + r_s cu \frac{dv}{d\xi} - r_s j(v) = 0$$

which becomes

$$\frac{d^2v}{d\xi^2} + r_s cu \frac{dv}{d\xi} = 0 \qquad \text{for} \quad v < V_1 \qquad (4.2.14a)$$

and

$$\frac{d^2v}{d\xi^2} + r_s cu \frac{dv}{d\xi} - r_s g_a(v - V_2) = 0 \qquad \text{for} \quad v > V_1 \qquad (4.2.14b)$$

If, for convenience, we choose $\xi = 0$ to be where $v = V_1$, a leading edge that makes a transition between zero and $V_2$ (see Fig. 4-6) and satisfies (4.2.14) is easily constructed.

Thus

$$v = V_1 e^{-\gamma_1 \xi} \qquad \text{for} \quad v < V_1 \qquad (4.2.15a)$$

and

$$v = V_2 - (V_2 - V_1) e^{\gamma_2 \xi} \qquad \text{for} \quad v > V_1 \qquad (4.2.15b)$$

where $\gamma_1 = r_s cu$ and $\gamma_2 = (r_s cu/2)[-1 + (1 + 4g_a/r_s c^2 u^2)^{\frac{1}{2}}]$. The velocity of propagation is not yet determined in (4.2.15) but may be computed in either of two ways (Scott, 1962): (a) by equating the total power being produced by $j(v)$ and absorbed by $r_s$ over the waveform to $\frac{1}{2} cV_2^2 u$, the power being absorbed by the membrane capacitance at velocity $u$, or (b) demanding continuity in the longitudinal current, $i$, at $\xi = 0$. Approach (a) is employment of the power balance idea behind (4.3). The leading edge must absorb energy (electrical energy in the membrane capacitance) at the same rate it is being produced for a steady traveling wave to exist. Approach (b) is equivalent to (a) and somewhat more convenient. From (4.2.15) and (4.2.4a), $i(\xi)$ is easily calculated for the ranges $\xi > 0$ and $\xi < 0$, and current continuity at $\xi = 0$ implies $\gamma_2 (V_2 - V_1) = \gamma_1 V_1$, which is readily solved for the velocity as

$$u = \left[ \frac{g_a}{r_s c^2} \right]^{\frac{1}{2}} \frac{(V_2 - V_1)}{(V_2\, V_1)^{\frac{1}{2}}} \qquad (4.2.16)$$

The case $g_r \neq 0$ has been studied in detail by Kunov (1966) and by Vorontsov, Kozhevnikova, and Polyakov (1967), who find

$$u = \frac{\left[ \left( \frac{V_2 - V_1}{V_1} \right)^2 g_a - g_r \right]}{\left[ r_s c^2 \frac{V_2 (V_2 - V_1)}{v_1^2} \left( \frac{V_2 - V_1}{V_1} g_a + g_r \right) \right]^{\frac{1}{2}}} \qquad (4.2.17)$$

(see also Kompaneyets, 1971). It can be seen that a necessary condition for a steady wave of transition from $v = 0$ to $V_2$ is $(V_2 - V_1)^2 g_a > V_1^2 g_r$. This again implies again that the areas $A_2$ and $A_1$ in Fig. 4-8 must satisfy the inequality

$$A_2 > A_1 \qquad (4.2.18)$$

The effect of "narcotization", discussed in the previous section in connection with Fig. 4-4, is to reduce $g_a$. Eventually the inequality (4.2.18) is violated and only decremental conductance can take place.

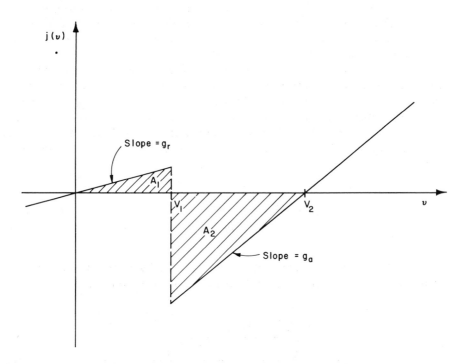

FIGURE 4-8.    "Piecewise linear" approximation for $j(v)$ (Tasaki, 1968).

The value of (4.2.16) can be assessed by using it to calculate the velocity of the action potential for the Hodgkin-Huxley axon shown in Fig. 4-2.    From the Hodgkin-Huxley axon parameters given on page 69 and taking $G_a$ = 33 mmho/cm$^2$ from Fig. 4. 2, the factor $(g_a/r_s c^2)^{\frac{1}{2}}$ is equal to 33. 7 m/sec.    Taking $V_2 = V_{Na} - V_R$ = 115 mV and (from Fig. 3-10) $V_1$ = 30 mV gives u = 48 m/sec, which is more than a factor of two higher than that calculated by Hodgkin and Huxley.    The main source of this error is the assumption made at the beginning of this section that $\tau_m$ = 0.    This assumption implies that sodium current will begin to flow fully as soon as the membrane voltage changes by 30 mV.    But inspection of Figs. 3-12 or 4-2 indicates that this is not so.    The time delay associated with sodium turn on requires the membrane voltage to change by about 60 mV before the membrane conductance rises to half of its full active value.    Taking $V_1$ = 60 mV gives

$$u = 22 \text{ m/sec}$$

which is satisfactory considering the nature of the approximations that have been made.

A more flexible procedure for taking account of the sodium turn on delay has recently been developed by Rissman (1977).    He assumes $T(v)$ defined in (4. 2. 7) to have the form

$$T(v) = A \sin \frac{\pi v}{V_2} \qquad (4. 2. 19)$$

which ensures a transition wave of amplitude $V_2$ as indicated in Fig. 4-6.    Then from (4. 2. 8) the form of $j(v)$ can easily be computed as

$$j(v) = K_1 \sin \frac{\pi v}{V_2} + K_2 \sin \frac{2\pi v}{V_2} \qquad (4. 2. 20)$$

where the adjustable constants $K_1$ and $K_2$ are related to the unknown constants A and u by

$$A = \sqrt{\frac{2r_s V_2 K_2}{\pi}} \quad \text{and} \quad u = \sqrt{\frac{\pi K_1^2}{2K_2 r_s c^2 V_2}} \qquad (4. 2. 21a,b)$$

Now $K_1$ and $K_2$ can conveniently be chosen to fit (4. 2. 20) to the experimental measurement of ion current plotted against membrane voltage as obtained from Figs. 1-3 or 4-2.   Rissman finds in general that the ratio

$$(K_2/K_1) = 0.6$$

gives good fit to such data.   Their magnitudes, of course must be adjusted to account for the axon circumference.   For the Hodgkin-Huxley axon with a maximum ion current per unit length of $1.07 \times 10^{-2}$ A/m  he calculates

$$u = 23.7 \text{ m/sec}$$

A somewhat similar (piecewise linear) model for $j(v)$ has recently been studied by Pastushenko, Chizmedzhev and Markin (1975a).

Donati and Kunov (1976) have discussed the application of velocity formula (4.2.17) to measurements on eight squid giant axons.   They assumed $V_2 = V_{Na} - V_R$, and took $V_1$ to be at the point of maximum rise on the leading edge.   Values for $V_1$ ranged between 52 mV and 69 mV with an average of 57 mV.   Since the active state conductance, $g_a$, is difficult to measure, this equation is not particularly useful for an absolute estimate of the conduction velocity.   However, it is quantitatively useful in predicting small changes in velocity due to changes in such parameters as external ionic concentrations, temperature, and drug content. Donati and Kunov, for example, use calculations of the change of membrane conductance in the wake of a pulse together with (4.2.17) to predict the conduction velocity of a second pulse.   The agreement that they display between predicted and measured velocity changes is quite impressive.

The importance of time delay in the conductance rise was emphasized by Offner, Weinberg, and Young (1940), who developed a velocity formula similar to (4.2.16) shortly after Cole and Curtis (1939) recorded the waveforms displayed in Fig. 1-3.   This delay is also of theoretical importance since (4.2.16) and (4.2.17) imply

$$u \rightarrow \left[ \frac{g_a}{r_s c^2} \frac{V_2}{V_1} \right]^{\frac{1}{2}} \rightarrow \infty \text{ as } V_1 \rightarrow 0 \qquad (4.2.22)$$

but with $\tau_m \neq 0$, the effective value of $V_1$ cannot reach zero.

Thus an infinite propagation velocity is prevented by the nonzero value of $\tau_m$.

Early attempts to calculate the propagation velocity of an action potential have been reviewed by Offner, Weinberg, and Young (1940). Since that time, additional approaches have been developed by Rosenblueth, Wiener, Pitts, et al. (1948), Huxley (1959), Scott (1962), Kompaneyets and Gurovich (1965), Balakhovskii (1968), Namerow and Kappl (1969), Smolyaninov (1969), Pickard (1966), and Markin and Chizmadzhev (1967), of which the last two references relate propagation velocity to the rate of rise on the leading edge of the action potential. Such a relation is easily obtained from (6.15a) since

$$\frac{\partial v}{\partial t}\bigg|_{max} = -u \frac{dv}{d\xi}\bigg|_{\xi=0} = \gamma_1 u V_1$$

Thus

$$u = \left| \frac{v_{t,max}}{r_s c V_1} \right|^{\frac{1}{2}} \tag{4.2.23}$$

as is readily verified for the waveform in Fig. 4-2. This is the formula used by Zeeman (1972).

Quite recently Pastushenko, Chizmedzhev and Markin (1975b) have indicated how sodium turn-on delay can be included in a velocity formula. They continue to assume $\tau_n = \tau_h = 0$ but approximate $m_0(v)$ as a step function and take $\tau_m = $ const. $(\neq 0)$. Then they integrate the third order system $(v, i, m)$ to find

$$u = \gamma \left[ \frac{g_{Na}}{r_s c^2} \right]^{\frac{1}{2}} \left[ \frac{h_0(V_2 - V_1)}{V_1} \right]^{\frac{1}{2}} \tag{*}$$

where $\gamma$ is the positive real root of

$$\gamma^2 (\gamma^2 + a)(\gamma^2 + 2a)(\gamma^2 + 3a) = 6a^3$$

and

$$a \equiv \frac{c V_1}{\tau_m g_{Na} h_0 (V_2 - V_1)}$$

Note that in the limit $\tau_m \to 0$, $\gamma \to 1$ and the velocity formula (*) becomes similar to (4. 2. 16). For the Hodgkin-Huxley axon at 18. 5°C, they take $V_2 = 115$ mV, $V_1 = 30$ mV, $\tau_m = 1.31 \times 10^{-4}$ sec and $\bar{G}_{Na} = .12$ mho/cm$^2$ (see Fig. 3-12a) together with the parameters listed on page 69. Then $a = .045$ and

$$\gamma = .295 \quad \text{so} \quad u = 19 \text{ m/sec}$$

We now turn our attention briefly to the effect of magnetic fields, which are associated with the longitudinal currents and represented as the inductors $\ell_i + \ell_o = \ell_s$ in Fig. 2-4b, on the propagation velocity. This question arises because it has been suggested (Lieberstein, 1967a, b, 1973; Brady 1970, Isaacs 1970, Lieberstein and Mahrous 1970, Lieberstein, 1973) that (2. 30) could be augmented to the form[*]

$$\frac{\partial^2 v}{\partial x^2} - \ell_s c \frac{\partial^2 v}{\partial t^2} = r_s \left( c \frac{\partial v}{\partial t} + j_i \right) + \ell_s \frac{\partial j_i}{\partial t} \qquad (4. 2. 24)$$

Then they erroneously assume that the numerical instability discussed in connection with Fig. 4-1 is related to physical instability, and they set both sides of (4. 2. 24) to zero at a velocity

$$u = [\ell_s c]^{-\frac{1}{2}} \qquad (4. 2. 25)$$

To examine this question (see Scott, 1971b) we again ignore turn-on delay and assume $\tau_m = 0$. The first-order PDEs corresponding to Fig. 2-4b and (4. 2. 24) become

$$v_x = -\ell_s i_t - r_s i$$

$$(4. 2. 26)$$

$$i_x = -cv_t - j(v)$$

Taking $j(v)$ as in Fig. 4-8 with $g_r = 0$ and assuming a steady wave of propagation, $v(x-ut) = v(\xi)$, then yields (Scott, 1963, 1970)

---

[*] Van Der Pol (1957) has proposed a similar model for propagation on a nerve fiber.

$$u = \left\{ \frac{\left( \dfrac{g_a}{r_s c^2} \right) \left[ \dfrac{(V_2 - V_1)^2}{V_1 V_2} \right]}{\left[ 1 - \dfrac{g_a \ell_s}{r_s c} \dfrac{(V_2 - V_1)}{V_2} \right]} \right\}^{\frac{1}{2}}$$

(4. 2. 27)

This implies that series inductance will have a negligible effect on velocity if it satisfies the inequality

$$\ell_s << \left( \frac{r_s c}{g_a} \right) \left( \frac{V_2}{V_2 - V_1} \right)$$

(4. 2. 28)

The left-hand side of (4. 2. 28) can be evaluated from (2. 12) and (2. 14), using small argument approximations for the Bessel functions, as

$$z_1 + z_2 \approx \frac{1}{\pi \sigma_1^* a^2} + i\omega \frac{\mu_0}{4\pi} [1 - 2 \log (\beta a)]$$

(4. 2. 29)

the second term of which gives the series reactance from magnetic fields both inside and outside the fiber.  Thus

$$\ell_s = \frac{\mu_0}{4\pi} [1 - 2 \log (\beta a)]$$

where $\mu_0$ $(= 4\pi \times 10^{-7}$ H/m) is the MKS magnetic permeability of nonmagnetic materials.  Taking $\beta a \sim 10^{-2}$ implies $\ell_s \sim 10^{-6}$ H/m.  The right-hand side of (4. 2. 28) is greater than 100 H/m; thus the inequality is satisfied by eight orders of magnitude and magnetic energy storage will have no measurable effect on the normal propagation of an action potential.  This conclusion is further supported by the numerical studies of Kaplan and Trujillo (1970).  Solutions of (4. 2. 24) at the velocity given in (4. 2. 25) for which both sides of the equation go to zero represent nothing more than a decoupling of high-frequency electromagnetic waves from the membrane.  Although this may have been what Newton (1718) had in mind when he posed his "twenty-fourth question, " it does not correspond to normal nerve activity.

## 3.   THE FITZHUGH-NAGUMO EQUATION

Sections 4-1 and 4-2 have bracketed (in the sense of an artilleryman) the representation of a propagating nerve fiber.   The Hodgkin-Huxley equations (4.1.1) and (4.1.2), give a fairly accurate description of spike propagation but are somewhat difficult to analyze without the aid of an automatic computer.   The nonlinear diffusion equation (4.2.2) is simple enough for analytical investigation and yields some useful results [e.g., (4.2.17) for the conduction velocity], but it fails to reproduce the qualitatively important feature of pulse recovery that is necessary for repeated firing of the fiber.   In this situation FitzHugh (1961) and Nagumo, Arimoto, and Yoshizawa (1962) proposed a modification of the nonlinear diffusion equation that would retain its simplicity but allow the action potential to return to a resting level.   In properly chosen units of space, time, and voltage, (4.2.2) can be written $V_{xx} - V_t = F(V)$, where $F(V)$ is a function with the character indicated in Fig. 4-6a or 4-8.   Augmenting this equation with a new "recovery" variable $R$ to (FitzHugh, 1969)

$$V_{xx} - V_t = F(V) + R \qquad (4.3.1a, b)$$

where

$$R_t = \epsilon (V + a - bR)$$

yields the desired effect.   To see this note that $R$ in (4.3.1a) acts as an outward ion current tending to decrease the area $A_2$ in Figs. 4-6a or 4-8.   With reference to the Hodgkin-Huxley equations (4.1.1) and (3.2.4a), there is a correspondence between

$$R \sim n$$

$$\epsilon b \sim \kappa \tau_n^{-1}$$

$$\epsilon V \sim \kappa n_o \tau_n^{-1}$$

where $\kappa$ is the "temperature factor" indicated in (3.2.6).   The constant $a$ in (4.3.-1b) can be absorbed into the definition of $R$ and $F$ so there is no loss of generality in setting it to zero.   The constant $b$ is often arbitrarily assumed equal to zero.   Since $\epsilon$

is proportional to $\kappa$, it can be considered as a parameter that increases with temperature.

Equations (4. 3. 1) are beginning to assume the role with respect to nerve-fiber propagation that the equation of Van Der Pol (1926, 1934) has played with respect to oscillator theory. "Van Der Pol's equation" displays the qualitative features of many oscillators (spontaneous excitation, limit cycle, continuous transition between sinusoidal and blocking behavior, etc. ) without necessarily being an exact representation of any particular dynamical system. As recent studies (Cohen, 1971; Hastings, 1972; 1976a), Greenberg, 1973) indicate, such a model is very stimulating and useful for the applied mathematician. Equations (4. 3. 1) are often called "Nagumo's equation" (McKean, 1970; Greenberg, 1973), although FitzHugh (1969) refers to it as the "B. V. P. equation" in recognition of the introduction by Bonhoeffer (1948) of phase-plane analysis to study the passive-iron nerve model, and of Van Der Pol. The reference to Van Der Pol, however, is somewhat unfortunate for in 1957 he introduced his own modification for application to nerve problems that failed to emphasize the diffusive character of a nerve fiber. Thus the term "FitzHugh-Nagumo equation" assigned by Cohen (1971), Rinzel and Keller (1973), and Hastings (1976b) seems most appropriate.

The general utility of (4. 3. 1) can be appreciated by considering the design of a neuristor or electronic analog of the active nerve fiber proposed by Crane (1962). Equations (4. 3. 1) describe the most natural technique for achieving pulse return in an electronic neuristor (Nagumo, Arimoto, and Yoshizawa, 1962; Crane, 1962; Scott, 1962, 1964; Berestovskii, 1963; Noguchi, Kumagai, and Oizumi, 1963; Yoshizawa and Nagumo, 1964; Sato and Miyamoto, 1967) and are closely related to the dynamical equations for active superconducting transmission lines that employ tunneling of either normal electrons (Giaever type) or superconducting electrons (Josephson type) (Scott,1964, 1970; Parmentier, 1969, 1970; Johnson, 1968; Nakajima, Yamashita, and Onodera, 1974; Nakajima, Onodera, Nakamura, and Sato, 1974; Reible and Scott,1975; Nakajima, Onodera, and Ogawa, 1976). Considered as a model for the nerve axon, (4. 3. 1) neglects: (a) turn-on delay for the sodium current, (b) the fourth-power dependence of potassium current on n, and (c) the dependence of $\tau_n$ on v. More exact second-order systems have recently been considered by Krinskii and Kokoz (1973). A good general survey of these problems is

given in the thesis by Kunov (1966), and a particular example is
presented in Section 4-8.

The analysis of (4. 3. 1) was begun by Nagumo, Arimoto, and
Yoshizawa (1962), who considered the ordinary differential equations for traveling-wave solutions of the form $V = V(x - ut) = V(\xi)$
and $R = R(x - ut) = R(\xi)$ as indicated in (4. 1. 3). Thus V and R
must then satisfy

$$\frac{dV}{d\xi} = W$$

$$\frac{dW}{d\xi} = F(V) + R - uW$$    (4. 3. 2a, b, c)

$$\frac{dR}{d\xi} = \frac{\epsilon}{u}(bR - V - a)$$

Nagumo and coworkers assumed $F(V)$ to be cubic, took $b = 0$
and obtained numerical evidence for the existence of two homoclinic trajectories for sufficiently small values of $\epsilon$. At a critical value, $\epsilon_c$, these solutions merged and for $\epsilon > \epsilon_c$ no homoclinic orbits were found, just as in Fig. 4-4. Such results sug-
gest the existence of two pulse-like traveling-wave solutions to
(4. 3. 1) (as in Fig. 4-3), and experiments on an electronic analog
indicated only the pulse with higher velocity to be stable. These
results were confirmed by FitzHugh (1969) through numerical studies of (4. 3. 1) and (4. 3. 2) with $b \neq 0$ and

$$F(V) = \frac{1}{3}V^3 - V$$    (4. 3. 3)

Velocities of the two branches are plotted against the "temperature parameter" $\epsilon$ in Fig. 4-9. FitzHugh also made a motion
picture entitled "Impulse propagation in a nerve fiber"[*] based on
numerical integration of (4. 3. 1). Some selected frames from this
film are reproduced in Fig. 4-10, which shows the propagation of
two pulses away from a point of stimulation. In the fully developed pulses (Figs. 4-10f-h) the recovery variable, R, follows behind
the voltage, V. These pulses correspond to the upper velocity (A')

---

[*] Available on loan from the National Medical Audiovisual Center
(Annex) Station K, Atlanta, Georgia 30333.

at $\epsilon$ = 0.08 in Fig. 4-9.  The lower velocity pulse (B') is unstable.  The locus of allowed traveling-wave velocities in the
$(u,\epsilon)$ plane indicates where the power balance condition (4.3) is
satisfied.  For $\epsilon > \epsilon_C$, only decremental conduction is possible.

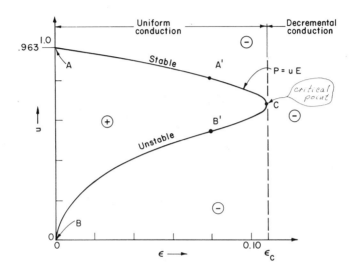

FIGURE 4-9.  Propagation velocity for traveling wave pulse solutions of the FitzHugh-Nagumo equations (4.3.1)
plotted against the "temperature parameter" $\epsilon$ for
a = 0.7 and b = 0.8 (redrawn from FitzHugh, 1969).

Arima and Hasegawa (1963) have considered a generalized form
of (4.3.1) with $R_t$ = G(V).  With suitable restrictions on F, G, and
the smoothness of the initial data, they show that a unique solu-
tion exists in the half space $|x| > \infty$ and $t > 0$.  Expanding this
result, Yamaguti (1963) showed that solutions of (4.3.1) with a =
0, b = 0, and VF $\geq$ CV$^2$ tend uniformly to zero.  A related result
was obtained by Yoshizawa and Kitada (1969), who consider (4.3.1)
with b = 0 but F(V) a cubic polynomial.  They confirm the existence of a threshold by showing that every solution in some neighborhood of zero converges to zero with increasing time.
The existence of homoclinic trajectories (which begin and end
at the same singular point) for (4.3.2) has been studied in detail

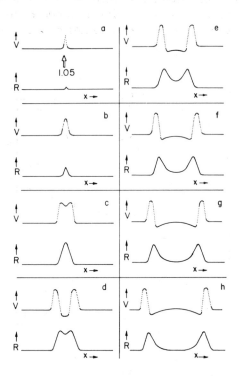

FIGURE 4-10.    Frames from the computer movie of Fitz Hugh showing results of a local stimulation of (4. 3. 1)  5%
above threshold with $\epsilon$ = 0.08,   a = 0.7, and  b = 0.8.

by Carpenter (1974, 1977a) and Hastings (1976b); and important analytical results have been obtained by Casten, Cohen, and Lagerstrom (1975) using singular perturbation methods.   Consider Fig. 4-
11, which shows two homoclinic orbits in the limit  $\epsilon$ = 0, implying (4. 3. 2c) R = const.   Orbit B corresponds to point B in Fig.
4-9 and is just the trajectory given in (4. 2. 6) for the zero velocity
"threshold pulse" shown in Fig. 4-7.   Orbit A, which corresponds
to point A in Fig. 4-9, is somewhat more complex.   It is the singular orbit approached as  $\epsilon \rightarrow 0$  of a family of homoclinic orbits
that correspond to the pulse shown in Fig. 4-12.   Going backward

in $\xi$, or forward in time, this pulse can be described as follows (encircled numbers correspond to branches in Figs. 4-11a and 4-12):

①   The "leading edge" involves a rapid transition between the outer zeros of $F(V)$ as was discussed in detail in the previous section.

②   A "slow relaxation" from $R = 0$ to a new value $R_a$ determined by (4.3.2c) with the condition $F(V) + R \approx 0$.

③   A rapid downward voltage transition between the two outer zeros of $F(V) + R_a$. The value of $R_a$ must be such that this trailing edge will have the same velocity as the leading edge (see Fig. 4-6).

④   Finally, there is a slow relaxation from $R = R_a$ back to zero.

The velocity, $u_0$, of the singular orbit A is just that velocity discussed in the previous section. Assuming $a = 0$, $b \neq 0$ and $\epsilon > 0$ we can write

$$u = u_0 + \epsilon u_1^A + \epsilon^2 u_2^A + \cdots$$

$$V = V_0 + \epsilon V_1 + \epsilon^2 V_2 + \cdots \qquad (4.3.4a, b, c)$$

$$R = R_0 + \epsilon R_1 + \epsilon^2 R_2 + \cdots$$

and equate powers of $\epsilon$ to obtain

$$\frac{d^2 V_0}{d\xi^2} + u_0 \frac{dV_0}{d\xi} - [F(V_0) + R_0] = 0$$

$$\frac{d^2 V_1}{d\xi^2} + u_0 \frac{dV_1}{d\xi} - V_1 F'(V_0) = R_1 - u_1^A \frac{dV_0}{d\xi} \qquad (4.3.5a, b, c)$$

$$\frac{dR_1}{d\xi} = \frac{bR_0 - V_0}{u_0}$$

$R_1(\xi)$ is readily obtained from integration of (4.3.5c). Then

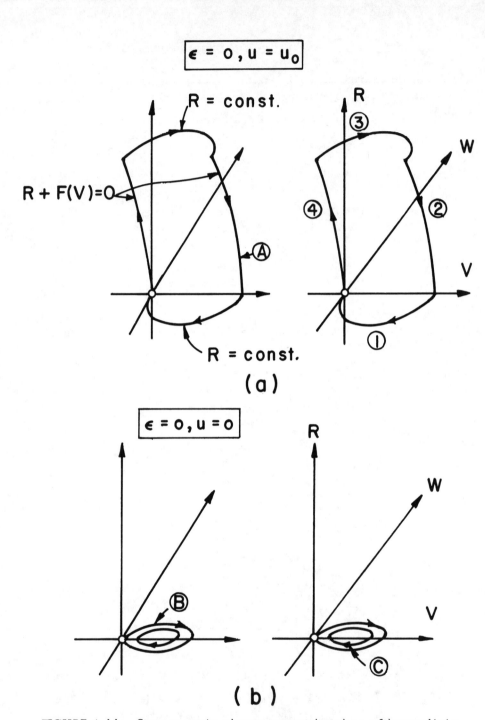

FIGURE 4-11. Stereoscopic phase space sketches of homoclinic trajectories for (4.3.2) with $\epsilon = 0$.
(a) $u = u_0 > 0$; (b) $u = 0$.

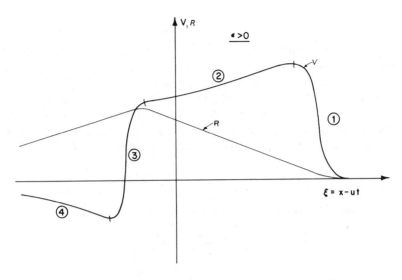

FIGURE 4-1 2.   Voltage pulse corresponding to orbit ④ in
Fig. 4-11a with $\epsilon > 0$.

$u_1^A$ can be determined in the following way.   Note that (4. 3. 5b)
can be written in compact form as

$$L V_1 = f \tag{4.3.5b'}$$

where $L \equiv (d^2/d\xi^2) + u_0 (d/d\xi) - F'(V_0)$ is a linear operator and
$f \equiv R_1 - u_1^A(dV_0/d\xi)$ is an inhomogeneous term.   Define an inner
product of two square integrable functions as

$$(v , w) = \int_{-\infty}^{\infty} v(\xi)w(\xi)\, d\xi$$

and recall that the adjoint $(\tilde{L})$ of an operator $L$ satisfies the
condition

$$(L v, w) = (v , \tilde{L}w) \tag{4.3.6}$$

Then the left-hand side of (4.3.5b) must be orthogonal to a function $\tilde{\phi}$ that satisfies

$$\tilde{L}\tilde{\phi} = 0 \qquad (4.3.7)$$

To prove this, check that $(LV_1, \tilde{\phi}) = (V_1, \tilde{L}\tilde{\phi}) = (V_1, 0) = 0$. From differentiation of (4.3.5a) it is seen that $\phi = (dV_0/d\xi)$ is a square integrable solution of the homogeneous equation $L\phi = 0$. Then $\tilde{\phi} = (dV_0/d\xi) \exp(u_0\xi)$ is easily shown to be a solution of the adjoint homogeneous equation (4.3.7). Thus the right-hand side of (4.3.5b) must also be orthogonal to $\tilde{\phi}$. This condition, $(\tilde{\phi}, f) = [\exp(u_0\xi) dV_0/d\xi), (R_1 - u_1{}^A dV_0/d\xi)] = 0$, determines $u_1{}^A$ as

$$u_1^A = \frac{1}{u_0} \frac{\int_{-\infty}^{\infty} \left[ \int_{\xi}^{\infty} (V_0(\xi') - bR_0(\xi'))d\xi' \right] \dfrac{dV_0}{d\xi} \exp(u_0\xi)d\xi}{\int_{-\infty}^{\infty} \left( \dfrac{dV_0}{d\xi} \right)^2 \exp(u_0\xi)d\xi} \qquad (4.3.8)$$

where it has been assumed in integrating (4.3.5c) that $R_1 \to 0$ as $\xi \to +\infty$. From Fig. 4-9 it is seen that the approximation

$$u \approx u_0 + u_1^A \epsilon \qquad (4.3.9)$$

is useful over a substantial portion of the upper (stable) branch.

   In computing $u_1^A$ from (4.3.8) it is convenient to choose $\xi = 0$ along the leading edge of the pulse (① in Figs. 4-11a and 4-12). The weighting function, $\exp(u_0\xi)$, then eliminates all contributions to the integrals from branches ②, ③, and ④ of the singular orbit. Furthermore, since $R_0 = 0$ along branch ①, the constant b [introduced in (4.3.1b)] does not enter into the calculation of $u_1^A$. As an example of the application of (4.3.8), take $F(v)$ to be as in Fig. 4-8 with $g_a = 1$ and $g_r = 0$ and note that the normalizations in (4.3.1) imply $r_s = 1$ and $c = 1$. Then, from (4.2.16), $u_0 = (V_2 - V_1)/(V_2V_1)^{\frac{1}{2}}$, and substitution of $V_0(\xi)$ from (4.2.15) into (4.3.8) gives (4.3.9) as

$$u = u_0 - \frac{2\epsilon}{u_0^3} \left[ 1 + \frac{V_1^2 - 3V_1V_2}{4V_2^2} \right] + O(\epsilon^2) \qquad (4.3.10)$$

This expansion is useful only if $\epsilon \ll u_0^4$.[*]

For the orbit Ⓑ in Fig. 4-11, $u_0 = 0$ and (4.39) cannot be used. In this case Casten, Cohen, and Lagerstrom (1975) write

$$V = V_0 + \sqrt{\epsilon}\, V_1 + \cdots$$

$$R = \sqrt{\epsilon}\, R_1 + \cdots \qquad (4.3.11a, b, c)$$

$$u = \sqrt{\epsilon}\, u_1^B + \cdots$$

to obtain

$$\frac{d^2 V_0}{d\xi^2} - F(V_0) = 0 \qquad (4.3.12)$$

$$\frac{d^2 V_1}{d\xi^2} - V_1 F'(V_0) = R_1 - u_1^B \frac{dV_0}{d\xi} \qquad (4.3.13)$$

$$\frac{dR_1}{d\xi} = \frac{-V_0}{u_1^B} \qquad (4.3.14)$$

The left-hand side of (4.3.13) is now orthogonal to $(dV_0/d\xi)$, so

$$u_1^B = + \left[ \frac{\int_{-\infty}^{\infty} V_0^2 \, d\xi}{\int_{-\infty}^{\infty} \left(\frac{dV_0}{d\xi}\right)^2 d\xi} \right]^{\frac{1}{2}} \qquad (4.3.15)$$

---

[*] Note that in (4.3.10) $V_1$ and $V_2$ are constant voltages defined in Fig. 4-8.

Again we see that the approximation

$$u \approx \sqrt{\epsilon} \; u_1^B \tag{4.3.16}$$

is useful over much of the lower branch in Fig. 4-9.

Closed trajectories satisfying (4.3.2) correspond to the periodic wave solutions originally suggested by Huxley (1959) for the Hodgkin-Huxley equations. The existence of such closed orbits for the FitzHugh-Nagumo equation has been studied by Hastings (1974) and by Carpenter (1974, 1977a) using the concept of "isolating blocks" (Conley and Easton, 1971) around a singular orbit.

Periodic solutions for the FitzHugh-Nagumo equation can be readily appreciated with reference to Fig. 4-13, which is closely related to Fig. 4-7. It is assumed that $R = 0$, $\epsilon = 0$ and $u = 0$ so (4.3.2a, b) imply

$$\frac{dV}{dW} = \frac{W}{F(V)} \tag{4.3.17}$$

which can be solved for the trajectory

$$W = [\, 2 \int^{V} F(V')dV'\,]^{\frac{1}{2}} \tag{4.3.18}$$

just as in (4.2.6). Then from (4.3.2a), $V(x)$ can be expressed as the elliptic integral

$$x = \int^{V_0} \frac{dV}{[\, 2 \int^{V} F(V')dV'\,]^{\frac{1}{2}}} \tag{4.3.19}$$

where simple zeros of

$$\int^{V} F(V')dV'$$

lead to turning points in the function $V_0(x)$ defined by (4.3.19). This situation is sketched in Fig. 4-13, where the necessary zeros are insured by equal positive and negative areas under $F(V)$ between the turning points. A closed trajectory corresponding to one of a family of such periodic solutions is indicated as C in Fig. 4-11b. For the case $\epsilon > 0$, the perturbation technique of Casten,

Cohen, and Lagerstrom (1975) can again be applied.  From equations (4. 3. 11) through (4. 3. 14) it is easily demonstrated that the left hand side of (4. 3. 13) is orthogonal to  $(dV_0/dx)$ , where the integration is over a period,  $\lambda$ , of  $V_0(x)$ .  Thus  u  varies with  $\epsilon$  as indicated in (4. 3. 16) with

$$u_1^B = + \left[ \frac{\int_0^\lambda V_0^2 \, dx}{\int_0^\lambda \left( \frac{dV_0}{dx} \right)^2 dx} \right] \qquad (4.3.20)$$

Rinzel and Keller (1973) have studied periodic solutions of (4. 3. 2) with  $\epsilon = .05$ ,  a = 0,  b = 0,  and

$$F(V) = V \qquad \text{for} \quad V < V_1$$
$$= V - 1 \quad \text{for} \quad V > V_1 \qquad (4.3.21)$$

This is the function of Fig. 4-8 with  $g_a = g_r = 1$  so the phase-space equations are linear except along the plane  $V = V_1$ .  For a periodicity defined by

$$V(\xi) = V(\xi + \lambda) \qquad (4.3.22)$$

some numerical values for velocity,  u,  and amplitude,  A,  are shown as functions of  $\lambda$  in Fig. 4-14.  There are two waves for each period, the slower wave being unstable.

Currently it is of great interest to extend such results to the full Hodgkin-Huxley equations (4. 1. 1) or to the corresponding ordinary differential equations for traveling-wave solutions (4. 1. 5). Evans and Shenk (1970) have shown that (4. 1. 1) has a unique solution for arbitrary bounded initial conditions with continuous dependence on the initial conditions.  Carpenter (1974, 1977a) has extended the concept of isolating blocks to the higher dimensional phase space associated with (4. 1. 5) and indicated in Fig. 4-1.  To do this she takes  m  to be a fast variable so that  v, i,  and  m  vary along branches ①  and ③  and only  n  and  h  vary along branches ②  and ④ .  The small parameters are then  $\tau_m, \tau_n^{-1}$ , and  $\tau_h^{-1}$ , and both homoclinic and periodic orbits are established.

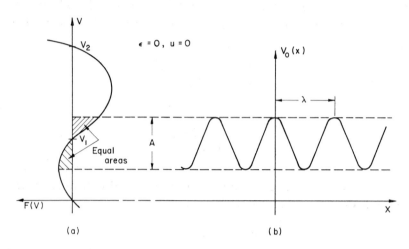

FIGURE 4-13.   Construction of periodic orbit   C    in Fig. 4-11b:
(a) "cubic" nonlinearity;  (b)  the periodic wave.

The analysis by Hastings (1976a), which does not assume $\tau_m$ to
be small, is probably closer to physiological reality as we saw in
the previous section.

Quite recently Carpenter (1977a, 1977b) has shown that the
Hodgkin-Huxley equations can exhibit traveling-wave solutions
that are qualitatively different from those of the FitzHugh-Nagumo
equation.   This arises because the Hodgkin-Huxley equations have
two slow variables (n and h) instead of one  (R)  for FitzHugh-
Nagumo.   These "new" solutions include:

1.  Finite pulse trains, where a fixed number $(1 < k < \infty)$ of
    pulses can propagate together.   In the phase space of
    Fig. 4-1a, this corresponds to a homoclinic orbit that
    "loops around" k times before returning to the singular point.

2.  Two periodic solutions for the same traveling-wave velo-
    city.   This differs from the FitzHugh-Nagumo case of Fig.
    4-14, which implies a single periodic solution for any
    particular value of the traveling-wave velocity.

3.  <u>Periodic bursting</u>, which is a periodic activity where the appearance of arbitrary numbers of pulses is separated by long intervals of rest.   More precisely, given any sequence of positive integers $\{k_i\}$, solutions of equations of Hodgkin-Huxley type can be constructed with $k_i$ pulses in the ith burst.

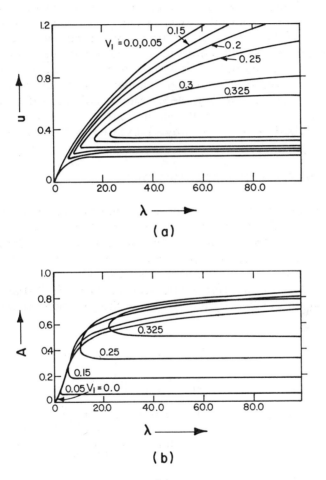

FIGURE 4-14.    (a) Velocity and (b) amplitude plotted against pulse spacing, $\lambda$, for periodic solutions of the FitzHugh-Nagumo equation with $\epsilon = 0.05$, $a = 0$, $b = 0$, and $F(V)$ as in (4. 3. 21) (redrawn from Rinzel and Keller, 1974).

There is, of course, no a priori reason to expect these exotic solutions to exist when $\tau_m, \tau_n^{-1}$, and $\tau_h^{-1}$ are set at values appropriate for a real nerve membrane.   However, the recent observations by Donati and Kunov (1976) indicate that such effects are at least plausible.   Donati and Kunov found that two pulses on a squid fiber become "locked together" when they are separated by a time interval of about 7 msec.   The reason for this effect can be appreciated from an examination of the structure of the stable Hodgkin-Huxley pulse (a) shown in Fig. 4-3.   On the wake of this pulse, the voltage $v_{12}$ passes from a value below $V_R$ (which would impede a trailing pulse) to a value above $V_R$ (which would accelerate a trailing pulse) at a time of about 7 msec.   This oscillation is, in turn, related to the "phenomenological inductance" discussed in relation to Fig. 4-5.   Such a locking effect should permit any number of Hodgkin-Huxley pulses to be hitched together (like railroad cars) in the ways indicated by Carpenter.

## 4.   THE MARKIN-CHIZMADZHEV MODEL

An interesting technique for analyzing the structure of a nerve pulse was introduced by Kompaneyets and Gurovich (1966) and discussed in detail by Markin and Chizmadzhev (1967).   This approach reproduces several of the qualitative features of the FitzHugh-nagumo equation but is even more simple.   Markin and Chizmadzhev start with the nonlinear diffusion equation (2.30) and immediately assume the ion current $j_i$ to be the following function of time

$$
\begin{aligned}
j_i(t) &= 0 &&\text{for}\quad t < 0 \\
&= -J_1 &&\text{for}\quad 0 < t < \tau_1 \\
&= +J_2 &&\text{for}\quad \tau_1 < t < \tau_1 + \tau_2 \\
&= 0 &&\text{for}\quad t > \tau_1 + \tau_2
\end{aligned}
\tag{4.4.1}
$$

Next the membrane voltage is taken to have the traveling-wave form $v = v(x - ut) = v(\xi)$.   Equation (2.30) is linear within each of the four regions indicated in Fig. 4-15.   General solutions within these regions take the forms:   ①   $A_1 \exp(-r_s \, cu \, \xi) + B_1,$

② $A_2 \exp(-r_s cu\xi) + B_2 - J_1\xi/cu$, ③ $A_3 \exp(-r_scu\xi) + B_3 + J_2\xi/cu$, and ④ $A_4 \exp(-r_s cu\xi) + B_4$. The requirement $v(\xi) \to 0$ as $|\xi|$ approaches infinity implies $B_1 = 0$ and $A_4 = 0$. Continuity of $v(\xi)$ and its first derivative at the boundaries between regions fixes the other six constants. The corresponding analytic expressions for $v(\xi)$ are:

Region ① :

$$v(\xi) = \frac{1}{r_s c^2 u^2}\{J_1 + J_2 \exp[-u^2 r_s c(\tau_1 + \tau_2)] - (J_1 + J_2)\exp(-u^2 r_s c\tau_1)\}\exp(-ur_s c\xi)$$

(4.4.2)

Region ② :

$$v(\xi) = \frac{1}{r_s c^2 u^2}\{J_2[\exp(-u^2 r_s c\tau_2) - 1] - J_1\} \exp[-r_s cu(\xi + u\tau_1)] - \frac{J_1\xi}{cu} + \frac{J_1}{r_s c^2 u^2}$$

(4.4.3)

Region ③ :

$$v(\xi) = \frac{J_2}{u^2 r_s c^2} \exp[-u^2 r_s c(\tau_1 + \tau_2) - ur_s c\xi] + \frac{J_2}{uc}\xi + \frac{\tau_1(J_1 + J_2)}{c} - \frac{J_2}{u^2 r_s c^2}$$

(4.4.4)

Region ④ :

$$v(\xi) = \frac{J_1\tau_1 - J_2\tau_2}{c}$$

(4.4.5)

From (4.4.5) the condition for pulse return to its starting value is

$$J_1\tau_1 = J_2\tau_2$$

(4.4.6)

(see Fig. 4-15).

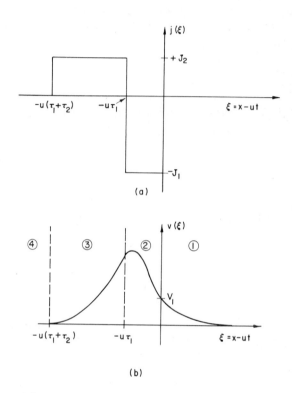

FIGURE 4-15.    (a) Membrane ionic current; (b) membrane voltage
assumed for a propagating pulse in the Markin-
Chizmadzhev model.


The velocity of propagation is not yet specified in (4.4.2)–
(4.4.4). It can be determined by the consistency condition that
at $\xi = 0$ (when inward ion current starts to flow) $v$ should equal
the threshold value $V_1$ as indicated in Figs. 4-6a or 4-8.    From
(4.4.2)

$$v(0) = \frac{1}{u^2 r_s c^2} \{J_1 + J_2 \exp[-u^2 r_s c(\tau_1 + \tau_2)] - (J_1 + J_2)\exp(-u^2 r_s c\tau_1)\}$$

$$(4.4.7)$$

which is sketched as a function of pulse velocity in Fig. 4-16.  It
is evident that, for a sufficiently small value of $V_1$, there are two
values of pulse velocity that satisfy the condition

$$V(0) = V_1 \tag{4.4.8}$$

If the upper value of pulse velocity, $u_A$, is slightly increased,
$v(0)$ becomes less than the critical value, $V_1$, and the pulse
slows down.  Thus this faster pulse appears stable.  In a similar
way, the lower value of pulse velocity, $u_B$, is for an unstable
pulse.  Thus the two intersections in Fig. 4-16 correspond to the
two branches of the velocity parameter curves in Figs. 4-4 and
4-9.

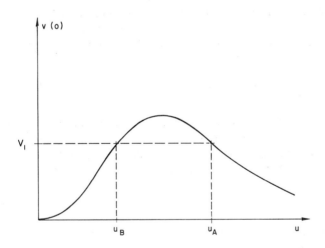

FIGURE 4-16.    Construction for stable $(u_A)$ and unstable $(u_B)$
pulse velocities in the Markin-Chizmadzhev model.

An approximate expression for the upper velocity, $u_A$, can be
determined under the assumption that $\tau_1$ is sufficiently large that
the exponential terms in (4.4.7) can be neglected.  Then

$$u_A \approx \sqrt{\frac{J_1}{V_1 r_s c^2}} \qquad (4.4.9)$$

Assuming, on the other hand, that the exponential terms in (4.4.7) are close to unity and also that (4.4.6) holds gives

$$u_B \approx \sqrt{\frac{2V_1}{r_s J_1 \tau_1 (\tau_1 + \tau_2)}}$$

The condition to be satisfied for both these approximations is

$$V_1 c \ll J_1 \tau_1 \qquad (4.4.10)$$

In a more recent paper, Undrovinas, Pastushenko, and Markin (1972) have indicated how membrane leakage current can be added to this model to account for the hyperpolarizing dip that appears at the end of the action potential in Figs. 1-3, 4-3, and 4-10.

## 5. THE MYELINATED AXON

Examination of (4.2.16) reveals a major design difficulty of the smooth nerve fiber. Since $g_a = 2\pi a G_a$, $c = 2\pi a C$, and $r_s = (\pi a^2 \sigma_1)^{-1}$, where $\underline{a}$ is the radius of the fiber, the conduction velocity is proportional to (Hodgkin, 1954)

$$u \propto a^{\frac{1}{2}} \qquad (4.5.1)$$

or to the fourth root of the cross-sectional area. [ See FitzHugh (1973) for a careful application of dimensional analysis to nerve problems. ] In order to double the velocity, the area (and hence the volume) of the fiber must increase by a factor of sixteen; to triple the velocity requires a factor of eighty-one. Since the giant axons of the squid transmit "escape signals" (generated in forward nerve cell complexes) to the appropriate muscles (located aft), there is evolutionary pressure to increase the speed, and this probably explains the unusually large size of the fiber. But clearly the fibers can't get much faster without occupying an

unacceptable fraction of the squid's cross-section, and only a
single bit of information ("leave" or "stay") is being transmitted
at any instant of time.    Equation (4.2.16) also indicates a solution to this dilemma.    If the fiber is partially covered by an insulating material so only a fraction, f, of the active membrane remains exposed, $(g_a/c)$ and $r_s$ would remain the same.    But c
would be proportional to f so the conduction velocity should depend on the exposed fraction roughly as

$$u \propto f^{-\frac{1}{2}}$$
(4.5.2)

Thus the velocity can be increased without changing the cross-
sectional area by making f small.

Something like this takes place in the design of the motor
axons of vertebrates.    The structure of the fiber appears as in
Fig. 4-17, where the fiber is covered almost everywhere by a relatively thick insulating coat of <u>myelin</u> consisting of a couple of
hundred layers of cell membrane (Hodgkin, 1951, 1964).    Only at
small active nodes (nodes of Ranvier) can the membrane function
in the normal way and these are spaced apart by a distance D ~
1 mm.    In this manner the diameter of the fiber can be as small as
$10 \mu$ while the conduction velocity is as large as that on the squid
fiber.    The frog nerve studied by Helmholtz (1850) and shown in
Fig. 1-1 is actually a bundle of many axons myelinated as in Fig.
4-17.    Young (1951) has prepared a graphic comparison of the squid
giant axon and the sciatic nerve of a rabbit (see Fig. 4-18).    The
conduction velocity is nearly idential in both cases, so the myelinated nerve bundle can carry at least two orders of magnitude
more bits of information per unit time.    This rapid information rate
permits the fine muscular control that is one of the striking features of higher animals.

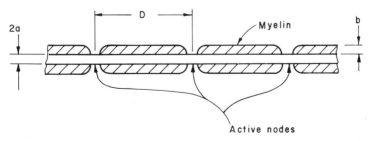

FIGURE 4-17.    Structure of a myelinated nerve fiber (not to scale).

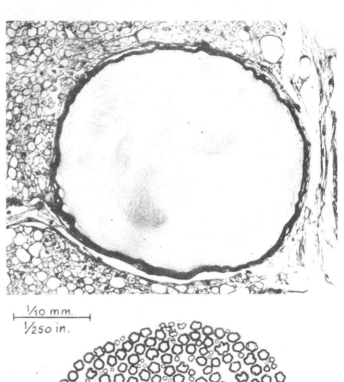

1/10 mm.
1/250 in.

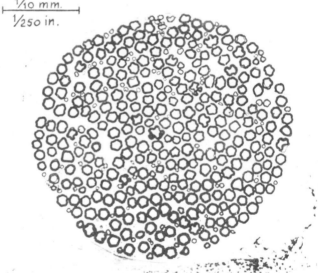

FIGURE 4-18.   Comparison of cross sections for the squid giant
axon (top) and the sciatic nerve bundle controlling
the calf muscle of a rabbit (below).   There are
about 400 myelinated fibers in the rabbit nerve,
each conducting pulses at about 80 m/sec.
(Young, 1951).  From Doubt and Certainty in Science
by J. Z. Young, published by Oxford University Press.

110

The role of isolated active nodes in increasing conduction speed was first recognized by Lillie (1925, 1936) in connection with his experiments on the "iron wire-acid" analog for the nerve fiber. He showed that the conduction velocity on this model was greatly increased when the wire was enclosed in a glass tube broken into segments, and he noted that the excitation seemed to "jump" quickly from one opening in the glass tubing to the next, an effect called "saltatory" conduction by physiologists. * Shortly thereafter Osterhout and Hill (1930) demonstrated that conduction in Nitella that had been blocked in fresh water by chloroform could be restored by introducing a salt bridge around the block, and Kato (1934) isolated in the conductable state a single fiber from the sciatic nerve of the Japanese toad. Building on these results, Tasaki (1939) demonstrated that conduction jumped from node to node in a single Japanese toad fiber. For general surveys of myelinated fibers the reviews by Tasaki (1959), Hodgkin (1951, 1964), and Waxman (1972) are suggested in addition to the discussion by Cole (1968). Here we list (Table 1) some representative data on the frog myelinated fiber collected by Hodgkin (1964).

TABLE 1.    Data on frog myelinated fiber

| | |
|---|---|
| Fiber radius  (a) | $7 \mu$ |
| Myelin thickness  (b) | $2 \mu$ |
| Distance between active nodes (D) | 2 mm |
| Area of active node | $2.2 \times 10^{-7} cm^2$ |
| Internal resistance per unit length | 140 M$\Omega$/cm |
| Capacity of myelin per unit length | 10-16 pf/cm |
| Conductance of myelin per unit length | $2.5 - 4 \times 10^{-8}$ mho/cm |
| Capacity of active node | 0.6 - 1.5 pF |
| Resting resistance of node | 40 - 80 M$\Omega$ |
| Conduction velocity | 23 m/sec |

---

* Saltare is the Latin and modern Italian verb "to jump. " For discussions of recent work on acid-wire nerve models, see Suzuki (1967) and Markin and Chizmadzhev (1974, Section 7. 1).

It is interesting to note how close his average conduction velocity is to the value of 27 m/sec measured by Helmholtz in 1850.

Equations (4.5.1) and (4.5.2) can be used to estimate the ratio of conduction velocity on the squid fiber $(u_s)$ to that of the frog fiber $(u_f)$. First of all, the square root of the (Hodgkin-Huxley) squid fiber radius $(238\mu)$ to the frog fiber radius in Table 1 $(7\mu)$ is 5.83. The fraction of exposed area, from Table 1 is $2.53 \times 10^{-4}$ so

$$f^{-\frac{1}{2}} = 62.9$$

Taking only these two factors into consideration would indicate

$$\frac{u_s}{u_f} = 0.09$$

whereas in fact the two velocities are about equal. One rather obvious additional correction is the extra internode (myelin) capacitance of the frog fiber. This increases the capacitance per unit length by the ratio

$$\frac{2.6}{.6} \quad \text{to} \quad \frac{4.7}{1.5} \quad \text{or} \quad 4.33 \quad \text{to} \quad 3.13$$

and, from (4.2.16) decreases the zero-order estimate of $u_f$ by the same factor. This gives an estimate of the velocity ratio in the range

$$\frac{u_s}{u_f} = 0.3 \quad \text{to} \quad 0.4$$

which is still substantially less than unity. There are various effects that might be invoked to explain this discrepency -- such as a difference in axoplasm resistivity or, from Fig. 3-12, a difference in membrane dynamics -- but one additional correction must necessarily be made. This is to account for the concentration of the active membrane at isolated points.

Such a correction can be effected by noting, from the considerations of Chapter 2, that the myelinated fiber is closely approximated by a linear diffusion equation that is periodically loaded by the active nodes (Pickard, 1966; Markin and Chizmadzhev, 1967).

This picture can be further simplified by lumping the internode capacitance of the myelin together with the nodal capacitance. This leads to the equivalent circuit indicated in Fig. 4-19, where

$$R = 28 \text{ M}\Omega$$

$$C = 2.6 - 4.7 \text{ pF}$$

and $I(k)$ is the ion current calculated at the $k$th node from (3.2.3) and (3.2.4) using the data in Fig. 3-12b. Equations (4.1.1a, b) are then replaced by the <u>difference differential equations</u> (DDEs)

$$v_k - v_{k-1} = -i_k R$$

$$i_{k+1} - i_k + C \frac{dv_k}{dt} = -I(k)$$
(4.5.3a, b)

which may also be written as second-order DDEs for the node voltages

$$v_{k+1} - 2v_k + v_{k-1} - RC \frac{dv_k}{dt} = RI(k) \qquad (4.5.3')$$

This is a DDE analog of the nonlinear diffusion equation (2.30). In order to determine a conduction velocity the traveling-wave assumption displayed in (4.1.3) and (4.1.4) must be replaced by a search for solutions that satisfy the condition

$$v_{k-1}(t) = v_k(t - T)$$

$$i_{k-1}(t) = i_k(t - T)$$
(4.5.4a, b)

where $T$ is a <u>section delay</u>. If $T$ can be found, the conduction velocity for the myelinated fiber is evidently

$$u_m = \frac{D}{T} \qquad (4.5.5)$$

In solving for the section delay, it is interesting to begin by assuming $I(k) = I(v_k)$, where

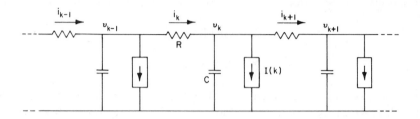

FIGURE 4-19.    Difference-differential representation of the myelinated nerve fiber.

$$I(v_k) = 0 \qquad \text{for} \quad v_k < V_1$$
$$= G(v_k - V_2) \quad \text{for} \quad v_k > V_1$$

(4. 5. 6)

as we did in (4. 2. 13) for the smooth axon.    Then (4. 5. 3') becomes a DDE analog of the K. P. P. equation (4. 1).    If conditions are such that $v_{k+1} - v_k \ll v_k$, the second difference in (4. 5. 3') can be approximated by a second derivative.    Then (4. 2. 16) can be used to calculate the section delay, $T_0$, as

$$\frac{T_0 G}{C} = \frac{[RG\, V_1 V_2]^{\frac{1}{2}}}{V_2 - V_1}$$

(4. 5. 7)

where we have used (4. 5. 5) and noted that $g_a \doteq (G/D)$, $r_s \doteq (R/D)$, and $c \doteq (C/D)$.    Since $(C/G)$ is the membrane time constant, the condition for validity of (4. 5. 7) is $(TG/C) \ll 1$ or $RG \ll 1$.    The problem is to determine $T$ as a function of $R, C, G, V_1$ and $V_2$ when this approximation is not valid.    In (4. 5. 3') time can be normalized to $RC$ and node voltage, to $V_2$.    Then the right-hand side can be written as a function of the two parameters, $RG$ and $(V_2 - V_1)/V_2$.    Thus the ratio of velocity on a myelinated axon, $u_m$, to the corresponding velocity on a smooth axon, $u_0 \equiv (D/T_0)$, can also be expressed as a function $\gamma(\cdot, \cdot)$ of the same two parameters

$$\frac{u_m}{u_0} = \frac{T_0}{T} = \gamma \left[ RG, \frac{V_2 - V_1}{V_2} \right] \tag{4.5.8}$$

The determination of this function was considered in detail by Kunov and Richer at the Electronics Laboratory of the Technical University of Denmark during 1964-65.  A detailed description of this work is included in the thesis by Kunov (1966) from which some of the salient points have been published (Kunov, 1965; Richer, 1965, 1966).  Kunov's thesis describes a variety of analytical studies including: (a) numerical integration of (4.5.3) for a finite number of sections, (b) an iterative computation to find solutions with the form (4.5.4), (c) a Laplace transform solution and (d) measurements on an electronic analog (Kunov, 1965). Their numerical results are summarized in Fig. 4-20.

For the frog axon the data in Fig. 3-12b give  $G = 0.57$ $\mu$mho so that $RG = 16$, and in Section 4-2 the value of $V_1$, which seemed to account for the delay in sodium turn-on, was about 60 mV. Thus $(V_2 - V_1)/V_2 = 0.5$.  From Fig. 4-20 these two values indicate a reduction in velocity of the myelinated fiber over that of a smooth axon by the factor  $\gamma = 0.4$, whereas our rough estimate obtained above by comparison of squid and frog fibers was  0.3 - 0.4.  This is a rather fortuitous agreement, considering the uncertainty in the capacitance  $C$  and the indication in Fig. 3-12b that the frog membrane responds somewhat more rapidly than that of the squid.  Furthermore, the appropriate value for  $G$  may not be as large as  0.57 $\mu$mhos since potassium and leakage currents flow in the opposite direction and, in addition, leakage current through the myelin and the resting conductance may have a noticeable effect as indicated in (4.2.17) (Kompaneyets, 1971).

Richer (1966) has made an important contribution to this problem by finding an exact solution for the case  $G = \infty$, which he calls "switch-line."  This solution gives an implicit relation between normalized section delay, $T_s/RC$, and  $(V_2 - V_1)/V_2$ as

$$\frac{V_2 - V_1}{V_2} = \exp \left[ - \int_0^{\pi/2} F(\alpha, \frac{T_s}{RC}) \operatorname{ctn}\alpha \, d\alpha \right] \tag{4.5.9}$$

where

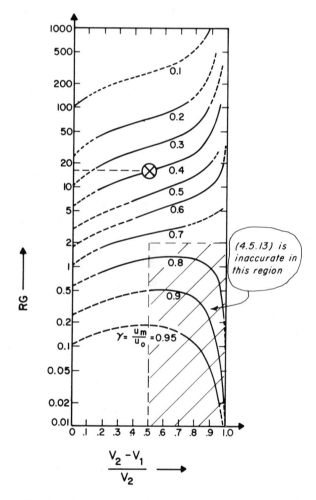

FIGURE 4-20.   Ratio of myelinated conduction velocity $(u_m)$ to
                that of the corresponding smooth fiber $(u_0)$ given
                by (4.2.16).   Dashed lines indicate extrapolated
                values (Kunov, 1966).

116

$$F(\alpha, \frac{T_s}{RC}) \equiv \frac{2}{\pi} \tan^{-1}[\,(ctn\alpha)\,\tanh\,(\frac{2T_s}{RC}\,\sin^2\alpha)]\qquad(4.5.10)$$

Equation (4. 5. 9) appears a bit unwieldy, but fortunately it can be approximated by the much simpler expression

$$\frac{T_s}{RC} \doteq \frac{V_1}{V_2 - V_1}\qquad(4.5.11)$$

which is found to be asymptotically correct for both large and small values of $T_s$ and overestimates $T_s$ by about 10% at $T_s/RC$ equal to unity.   Since (4. 5. 7) can be written in the form

$$\frac{T_0}{RC} = \frac{[V_1V_2/RG]^{\frac{1}{2}}}{V_2 - V_1}\qquad(4.5.7')$$

a simple interpolation between (4. 5. 11) and (4. 5. 7') is

$$\frac{T}{RC} \doteq \left(\frac{V_2}{V_2 - V_1}\right)\left[\left(\frac{V_1}{V_2}\right)\left(\frac{1}{RG} + \frac{V_1}{V_2}\right)\right]^{\frac{1}{2}}\qquad(4.5.12)$$

This equation agrees well with digital computer solutions for a long but finite system and also with the results of analog simulation (Kunov, 1966).   From (4. 5. 8) a simple approximation to the numerical curves in Fig. 4-20 is

$$\frac{u_m}{u_0} \doteq \frac{1}{\sqrt{1 + RG\,V_1/V_2}}\qquad(4.5.13)$$

This equation gives a fair representation of Fig. 4-20 in the range $RG > 1$. For example, with $RG = 16$ and $(V_2 - V_1)/V_2 = 0.5$ [or $(V_1/V_2) = 0.5$], (4. 5. 13) indicates $\gamma \doteq 1/3$, whereas Fig. 4-20 indicates $\gamma = 0.4$.   In the range $RG < 2$ and $(V_2 - V_1)/V_2 > 0.5$ [or $(V_1/V_2) < 0.5$], (4. 5. 13) is qualitatively incorrect. It indicates $\gamma \to 1$ as $V_1 \to 0$, whereas Fig. 4-20 indicates $\gamma \to 0$.

Richer (1965) has also considered the addition of resting conductance as in Fig. 4-8 and has shown that only a positive or negative level change can propagate (not a pulse) just as in the smooth axon. It is interesting to note that he finds an intermediate range for which neither wave can propagate.

Although the qualitative behavior of the function $\gamma(\cdot, \cdot) = (u_m/u_0)$ has been fairly well charted, an exact solution for the DDE (4.5.3') with $I(k) = I(v_k)$ has not yet been found. This problem remains as a challenge to applied mathematicians.

Kunov (1966) considers recovery models or discrete FitzHugh-Nagumo systems and Markin and Chizmadzhev (1967) discuss propagation when the internodes are described by the linear diffusion equation. FitzHugh (1962) computed the initiation and conduction of pulses on a linear diffusion equation periodically loaded with Hodgkin-Huxley nodes and improved computations have recently been reported by Goldman and Albus (1968).

The high velocity (stable) and low velocity (unstable) pulses that appear in Figs. 4-4 and 4-9 for the Hodgkin-Huxley and FitzHugh-Nagumo equations can be appreciated on the myelinated fiber by considering the "Nasonov diagram" (Averbach and Nasonov, 1950; FitzHugh, 1969) in Fig. 4-21. If it is assumed that: (a) each node has a "sigmoid" stimulus-response curve and (b) a fraction, $1/\alpha$, of the response for each node is presented as a stimulus to the next, then <u>stationary</u> levels of activity occur where the sigmoid curve intersects the line $R = \alpha S$. The lower amplitude intersection is unstable since a small increase in S leads to a larger increase in R, and so on. The upper intersection, on the other hand, appears to be stable. As the parameter $\alpha$ is increased, these two intersections eventually merge, and above this critical value of $\alpha$ only decremental conduction obtains.

At the beginning of this section it was noted that the conduction velocity for a smooth fiber is proportional to the square root of the fiber radius. Thus

$$u_s = k_1 a^{\frac{1}{2}} \tag{4.5.14a}$$

where for the Hodgkin-Huxley axon

$$k_1 = 1.43 \text{ m/sec } \mu^{\frac{1}{2}}$$

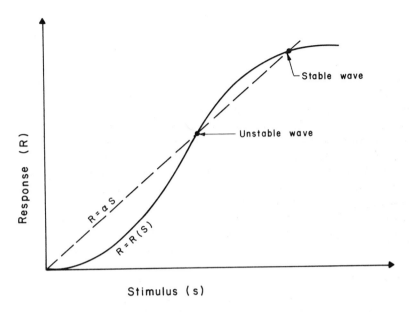

FIGURE 4-21.    Nasonov diagram for a myelinated nerve fiber.

For myelinated fibers there is evidence that conduction velocity is proportional to the radius itself.  Rushton (1951) explains this observation with the assumptions of:  (a) <u>constant nodal length</u>, which implies that $G \propto a$  and  $C \propto a$  and  (b) <u>internodal distance</u> (D) <u>proportional to radius</u>, which implies  $R \propto a^{-1}$.  Then from (4. 5. 12) the internodal delay (T) is independent of the radius; and from (4. 5. 5)

$$u_m = k_2 a \qquad (4.\,5.\,14b)$$

Using the frog fiber data of Table 1,

$$k_2 = 3.\,3 \text{ m/sec } \mu .$$

Rushton (1951) suggested that equations (4. 5. 14a, b) imply a critical radius  $(a_c)$  above which the conduction velocity will be faster for a myelinated fiber.  This critical radius is determined by setting $u_s = u_m$, where

$$a_c = \left(\frac{k_1}{k_2}\right)^2 = 0.18 \, \mu \qquad (4.5.15)$$

for the above mentioned values for $k_1$ and $k_2$. Waxman and Bennett (1972) have recently reexamined the data originally used by Rushton (1951) on small smooth and myelinated fibers of cats and kittens. They find evidence for a critical radius of $a_c \approx 0.1\mu$. Thus one might suppose that if speed of conduction is the important design criterion, then fibers with diameters less than 0.2 - 0.4 $\mu$ should not be myelinated. Waxman and Bennett suggest, however, that conduction speed may not be the only criterion in the mammalian central nervous system. Waxman, Pappas, and Bennett (1972) have found a functionally significant variability in the size and spacing of active nodes on the neural electric organ of the knife fish. In a study of the kangaroo rat's brain, Waxman and Melker (1971) observed certain fibers with ratios of internodal distance to diameter (including myelin) as low as 18:1. (For the frog fiber in Table 1 this ratio is 110:1.) Waxman and Melker indicate that the occurrence of closely spaced nodes in mammalian brain, and particularly in reticular formation, suggests that variations in the geometry of the central myelin sheath may provide a mechanism for "velocity matching" or more complex transformations of neural information in the axons of mammalian integrative neurons.

## 6.    FIBERS WITH CHANGING DIAMETER

In Chapter 2 we observed that if the dimensions of a fiber change slowly over a distance equal to the length of its action potential then (2.21) should be approximately correct with the local values of $r_s(x)$, $c(x)$ and $j_i(x, v, m, n, h)$. Such an approximation might, for example, be appropriate for calculations of pulse propagation on the large dendrites of the Mauthner cell of the goldfish (Bodian, 1952) shown in Fig. 4-22. Let us begin the study of gradual tapers by assuming (Lindgren and Buratti, 1969) that the parameters vary exponentially as

$$r_s = r_0 \, e^{\gamma x}$$

$$c = c_0 \, e^{-\gamma x} \qquad \text{(4.6.1a, b, c)}$$

$$j_i = j_0 \, e^{-\gamma x}$$

where $r_0$ and $c_0$ are constants and $j_0$ is independent of x. Then the nonlinear diffusion equation corresponding to (2.30) becomes

$$\frac{\partial^2 v}{\partial x^2} - \gamma \frac{\partial v}{\partial x} - r_0 c_0 \frac{\partial v}{\partial t} = r_0 j_0 (v, m, n, h) \qquad \text{(4.6.2)}$$

Using the Fitz Hugh-Nagumo model and normalizing as in (4.3.1), this equation can be approximated by

$$V_{xx} - \gamma V_x - V_t = F(V) + R$$

$$R_t = \epsilon (V - bR) \qquad \text{(4.6.3a, b)}$$

and the perturbation calculation of Casten, Cohen, and Lagerstrom (1975) can be carried through just as in Section 4-3. In particular, the pulse velocity is

$$u \approx u_0 + \gamma + u_1^A \epsilon \qquad \text{(4.6.4)}$$

where $u_0$ is the pulse velocity that would be calculated for R = 0 and $\gamma = 0$ using the concepts outlined in Section 4-2. The linear coefficient in $\epsilon$ becomes

$$u_1^A = \frac{1}{u_0 + \gamma} \cdot \frac{\displaystyle\int_{-\infty}^{\infty} \int_{\xi}^{\infty} \left[ \left( V_0(\xi') - bR_0(\xi') \right) d\xi' \right] \frac{dV_0}{d\xi} e^{u_0 \xi} \, d\xi}{\displaystyle\int_{-\infty}^{\infty} \left( \frac{dV_0}{d\xi} \right)^2 e^{u_0 \xi} \, d\xi}$$

$$\text{(4.6.5)}$$

which remains finite when $u_0 = 0$.  Thus (4. 6. 4) and (4. 6. 5) can be used for the slow wave as well as for the fast wave.   The primary effects of the (inward) exponential taper are to increase the zero-order term for pulse velocity in (4. 6. 4)(in the direction of increasing series resistance)and to decrease the magnitude of the first-order term.

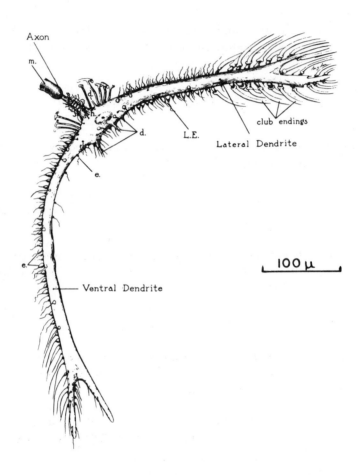

FIGURE 4-22.    Tapered dendrites of Mauthner's cell in the gold-
                fish; (d)  small dendrites; (e)  small endbulbs;
                (h)  axon hillock,  (LE) large endbulbs;
                (m)  myelin sheath on axon (Bodian, 1952).

The exponential variation assumed in (4. 6. 1) is not realistic if $r_s$ is almost entirely determined by the inside component and therefore inversely proportional to radius (a) squared as indicated in (2. 19). This is because both c and $j_i$ should be proportional to a. Equations (4. 6. 1) might be appropriate when the external current flow is restricted, as in (2. 25), and the dependence of $r_s$ on the radius is weakened.

Rall (1962a) has presented an interesting analysis of the tapered fiber that approaches an equation similar to (4. 6. 3) by transforming the spatial coordinate. He assumes internal resistance to dominate so that (2. 21a) becomes

$$i = - \pi a^2 \sigma_1 \frac{\partial v}{\partial x} \qquad (4.6.6)$$

and (2. 21b) becomes

$$N(v) = - [\frac{\partial i}{\partial x}][\frac{dA}{dx}]^{-1} \qquad (4.6.7)$$

where $(dA/dx)$ is the change in membrane area with increase in x. For a cylindrical membrane $(dA/dx) = 2\pi a$ just as in (2. 21b). But when the radius is a function of x, a differential application of the Pythagorean theorem implies

$$\frac{dA}{dx} = 2\pi a \left[ 1 + (\frac{da}{dx})^2 \right]^{\frac{1}{2}} \qquad (4.6.8)$$

Differentiating (4. 6. 6) with x and substituting into (4. 6. 7) then gives

$$\frac{2}{\sigma_1} N(v) = \frac{a}{\left[ 1 + (\frac{da}{dx})^2 \right]^{\frac{1}{2}}} \left[ \frac{\partial^2 v}{\partial x^2} + \frac{\partial v}{\partial x} (\frac{d}{dx} \log a^2) \right] \qquad (4.6.9)$$

At this point Rall defined a new spatial variable

$$z = z(x) \qquad (4.6.10)$$

so that

$$\frac{\partial v}{\partial x} = \frac{\partial v}{\partial z}\frac{dz}{dx}$$

and

$$\frac{\partial^2 v}{\partial x^2} = \left[\frac{dz}{dx}\right]^2\left[\frac{\partial^2 v}{\partial z^2} + \frac{\partial v}{\partial z}\frac{dx}{dz}\left(\frac{d}{dx}\log\frac{dz}{dx}\right)\right]$$

Substituting these derivatives into (4.6.9)

$$\frac{2}{\sigma_1}N(v) = \frac{a(\frac{dz}{dx})^2}{\left[1+(\frac{da}{dx})^2\right]^{\frac{1}{2}}}\left[\frac{\partial^2 v}{\partial z^2} + \frac{\partial v}{\partial z}\left(\frac{dx}{dz}\frac{d}{dx}\log(a^2\frac{dz}{dx})\right)\right] \qquad (4.6.11)$$

Then if the transformation $z(x)$ is chosen such that

$$\frac{dz}{dx} = a^{-\frac{1}{2}}\left[1 + (\frac{da}{dx})^2\right]^{\frac{1}{4}} \qquad (4.6.12)$$

(4.6.11) can be written

$$\frac{\partial^2 v}{\partial z^2} + K\frac{\partial v}{\partial z} = \frac{2}{\sigma_1}N(v) \qquad (4.6.13)$$

where

$$K \equiv \frac{\frac{d}{dx}[\log(a^2\frac{dz}{dx})]}{\frac{dz}{dx}} \qquad (4.6.14$$

$$= a^{\frac{1}{2}}\left[1 + (\frac{da}{dx})^2\right]^{-\frac{1}{4}}\frac{d}{dx}\log\left\{a^{3/2}\left[1 + (\frac{da}{dx})^2\right]^{\frac{1}{4}}\right\} \qquad (4.6.15)$$

Evidently $K = 0$ if $\underline{a}$ varies such that

$$a^{3/2} \left[ 1 + (\frac{da}{dx})^2 \right]^{\frac{1}{4}} = \text{const} \qquad (4.6.16)$$

and  K = const. if

$$z \propto \log (a^2 \frac{dz}{dx}) \qquad (4.6.17)$$

In a more recent analysis of tapered fibers, Goldstein and Rall (1974) assume the taper is sufficiently gradual so that

$$(\frac{da}{dx})^2 << 1 \qquad (4.6.18)$$

Then from (4.6.12)

$$\frac{dz}{dx} = a^{-\frac{1}{2}}$$

and (4.6.17) can be written

$$z = - \frac{1}{\gamma} \log (a^{3/2})$$

where  $\gamma$  is a constant.  Then

$$K = - \gamma \qquad (4.6.19)$$

and

$$a = (a_0^{\frac{1}{2}} - \frac{\gamma}{3} x)^2 \qquad (4.6.20)$$

where  $a_0$  is the fiber radius at  $x = 0$.

Thus for a radius variation as in (4.6.20) and assuming  $\gamma << 1$ so that (4.6.18) is satisfied, (4.6.13) is formally identical to (4.6.2) or (4.6.3) where  z  is related to  x  by

$$z = - \frac{3}{\gamma} \log (a_0^{\frac{1}{2}} - \frac{\gamma}{3} x) \qquad (4.6.21)$$

We can also suppose that the fiber diameter changes abruptly (see Fig. 4-23a).  In this case we should consider the effect of higher modes, which are necessary to match electromagnetic boundary conditions at the discontinuity.

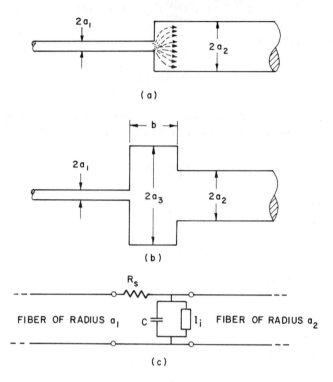

FIGURE 4-23.    (a)  Abrupt discontinuity;  (b)  varicose discontinuity;  (c)  lumped equivalent circuit for a discontinuity.

    To appreciate this problem remember that equations (2. 12) for
series impedance of the fiber were derived from (2. 8) which assume
the lowest order  TM  mode of propagation.   This mode is expected to dominate when the radius is changing slowly with  x,  but
generally,  there is an infinite number of such modes corresponding to higher order Bessel functions and more complex radial dependence.   The selected assortment of these higher modes which
match boundary conditions at the discontinuity will absorb energy
from the wave.   In microwave calculations the higher modes are
often represented by an equivalent admittance that appears at the
discontinuity [ see Ramo and Whinnery (1953) pp. 477-485 for an

introduction to these representations]. But for nerve fiber problems, where the distances are small compared with the electromagnetic penetration depth [see (4.15)], it can be assumed that
to a good approximation the electrical potential satisfies Laplace's
equation and the current "spreads" from the smaller to the larger
fiber as indicated in Fig. 4-23a. Then the discontinuity region
can be represented by a series <u>spreading resistance</u>

$$R_s \approx \frac{a_2 - a_1}{2\pi \sigma_1 a_1 a_2} \tag{4.6.22}$$

plus the shunting effect of the extra membrane of area

$$A = \pi(a_2^2 - a_1^2) \tag{4.6.23}$$

presented at the discontinuity. Thus the shunt elements in the
equivalent circuit of Fig. 4-23 are

$$I_i = j_i A \tag{4.6.24}$$

$$C = c A \tag{4.6.25}$$

The discontinuity in Fig. 4-23a can be considered as a special
case of the "varicosity" in Fig. 4-23b for which the equivalent
circuit still applies if

$$R_s \approx \frac{a_1 a_3 + a_2 a_3 - 2a_1 a_2}{2\pi \sigma_1 a_1 a_2 a_3} \tag{4.6.26}$$

and

$$A = \pi(2a_3^2 + 2a_3 b - a_1^2 - a_2^2) \tag{4.6.27}$$

for $\sigma_1 \sim 3$ mho/m and $a_1 = .238$ mm (the Hodgkin-Huxley axon),
the time constant

$$R_s C \sim (\frac{a_2}{a_1})^2 \times 10^{-6} \text{ sec}$$

The series spreading resistance should be negligible when $R_s C$ is less than $10^{-4}$ sec or for discontinuities in which $(a_2/a_1) < 10$. In this case it is not unreasonable to include the shunt elements with the description of the larger fiber and ignore nonuniformity at the discontinuity altogether. Under this assumption Markin and Pastushenko (1969) have obtained a simple and important boundary condition on pulse velocity. From (2.21a) continuity of axial current at the discontinuity implies

$$a_1^2 \frac{\partial v_-}{\partial x} = a_2^2 \frac{\partial v_+}{\partial x} \tag{4.6.28}$$

where the "+" and "−" subscripts refer to respective values just to the right and left of the discontinuity. Continuity of membrane voltage at the discontinuity implies

$$\frac{\partial v_-}{\partial t} = \frac{\partial v_+}{\partial t} \tag{4.6.29}$$

Defining pulse velocity, $u$, for a point of constant voltage amplitude as

$$u \equiv - \frac{\partial v/\partial t}{\partial v/\partial x} \tag{4.6.30}$$

implies

$$\frac{u_-}{a_1^2} = \frac{u_+}{a_2^2} \tag{4.6.31}$$

Thus pulse velocity should be discontinuous at the discontinuity. We noted at the beginning of Section 4-5 that pulse velocity on a smooth fiber is proportional to the square root of the radius. If we define $u_1$ ($u_2$) as the pulse velocity far to the left (right) of the discontinuity, then

$$\frac{u_1}{a_1^{\frac{1}{2}}} = \frac{u_2}{a_2^{\frac{1}{2}}} \tag{4.6.32}$$

Goldstein and Rall (1974) have recently presented some numerical studies that illustrate the relation between (4. 6. 31) and (4. 6. 32). Using a model that simulates the squid axon, they compute the "peak velocity", $u_p$, for a pulse in the vicinity of a discontinuity. As the pulse approaches a widening $(a_2/a_1 > 1)$ it slows down (Fig. 4-24a). In a qualitative way this is to be expected from our previous consideration of tapered fibers; $\gamma$ is negative for a widening taper that [through (4. 6. 4)] tends to reduce the velocity. As the pulse passes through the discontinuity at $x = 0$, its peak velocity jumps to a large value, as should be expected from (4. 6. 31) and eventually falls back to the value required by (4. 6. 32). Note, however, that the discontinuity at $x = 0$ shown in Fig. 4-24 does not exactly match that implied by (4. 6. 31). This may be because (4. 6. 31) was derived for a point of constant voltage amplitude, whereas the data of Fig. 4-24 were computed for the peak velocity. Similar results were obtained by Berkinblit, Vvedenskaya, Gnedenko, et al. (1970) for an abrupt widening of the Hodgkin-Huxley fiber. For a five fold widening the velocity of the peak of the action potential falls from 19 m/sec to about 2 m/sec and then rises sharply to about 54 m/sec as the action potential passes through the discontinuity.

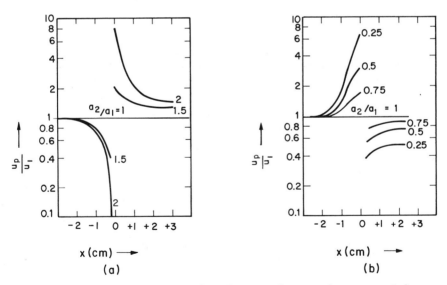

FIGURE 4-24.  Change of peak velocity of an action potential as it approaches: (a) abrupt enlargement; (b) abrupt construction (redrawn from Goldstein and Rall, 1974).

In Fig. 4-24b are plotted some corresponding data for a pulse that approaches a constriction $(a_2/a_1 < 1)$. Here the pulse gains in velocity as it approaches the discontinuity, which is to be expected since $\gamma$ is positive for a constricting taper. Then, as (4. 6. 31) implies, the pulse velocity falls rapidly upon passing through the discontinuity and eventually rises to the value required by (4. 6. 32).

Khodorov, Timin, Vilenkin, et al. (1969) have studied the propagation on a Hodgkin-Huxley axon that was abruptly increased in radius by factors of three, five, six, and ten. For the five-fold increase, an action potential did propagate through the discontinuity but with considerable delay ($\sim 0.8$ m/sec). With a six-fold increase, conduction was blocked. Waveforms for the case of marginal conductance are displayed in Fig. 4-25, which shows the enlarged portion of the fiber to initially propagate a pulse that is close to the threshold pulse (Fig. 4-3). This low-amplitude initial pulse eventually increases into a fully developed action potential. Berkinblit, Vvedenskaya, Gnedenko, et al. (1970) have extended this study by considering the effect of a tapered widening on the conditions for pulse blocking. Again the initial pulse was propagating on a Hodgkin-Huxley axon that widened over a taper distance an n-fold increase in radius. Their numerical observations are indicated in Table 2.

TABLE 2.   Critical diameter ratio as a function of taper length
(Berkinblit, Vvedenskaya, Gnedenko, et al., 1970)

| Length of Taper (cm) | Diameter Ratio | |
|---|---|---|
| | Blocking | Passage |
| 0. 088 | 5. 5:1 | 5:1 |
| 0. 785 | 6:1 | 5. 5:1 |
| 1. 76 | 8:1 | 7:1 |
| 3. 81 | >10:1 | 10:1 |

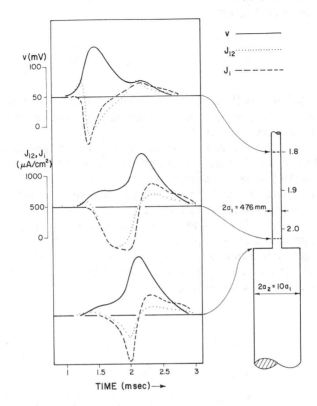

FIGURE 4-25.  Membrane voltage (v), total membrane current
density $(J_{12})$, and ionic current density $(J_i)$ for
the Hodgkin-Huxley action potential approaching
an abrupt five-fold enlargement (redrawn from
Khodorov, Timin, Vilenkin, et al., 1969).


Pastushenko and Markin (1969) have used the simple nerve
model of Section 4-4 to estimate the diameter ratio at which blocking of an action potential is observed.   A critical parameter in this
development is the ratio of maximum pulse voltage, $V_{max}$, to the
threshold voltage, $V_1$.   For large values of this <u>stability factor</u>

$$\kappa \equiv \frac{V_{max}}{V_1} \qquad\qquad (4.6.33)$$

they find that the condition for the enlarged fiber to be brought a-
bove its threshold level is

$$\left(\frac{a_2}{a_1}\right)^{3/2} < \kappa + 1.11\kappa^{\frac{1}{2}} - 1.69 \qquad (4.6.34)$$

From the stable and threshold waveforms for the Hodgkin-Huxley
axon shown in Fig. 4-3, $\kappa \approx 5$ which implies blocking for a diameter ratio greater than 3.22.

    Khodorov, Timin, Pozin, et al. (1971) have also considered the
propagation of a pulse train through an abrupt widening of the
Hodgkin-Huxley axon. Some examples of their computations are
presented in Fig. 4-26. For an abrupt widening of 5:1 and an incoming temporal period of 2.5 msec, only the first pulse will pass
(see Fig. 4-26a). If the widening ratio is reduced to 3:1 and the
temporal period is increased to 3.3 msec the fifth pulse will fail
but the sixth will again pass (see Fig. 4-26b). Calculations of
this sort are reviewed in the recent book by Khodorov (1974) and
can be summarized as follows.

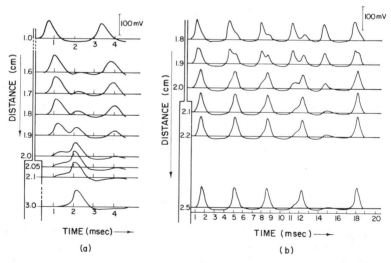

FIGURE 4-26.    Propagation of a pulse train through an abrupt widening of the Hodgkin-Huxley axon: (a) widening of
5:1 with an incoming temporal period of 2.5 msecs;
(b) widening of 3:1 with a period of 3.3 msecs (redrawn from Khodorov, Timin, Pozin, and Shmelev, 1971).

TABLE 3.    Blocking of pulses in a periodic train (Khodorov, 1974)

| Abrupt widening ratio | Temporal period (msec) | | | |
|---|---|---|---|---|
| | 2. 5 | 3 | 3. 3 | 3. 5 |
| | Number of blocked pulses | | | |
| 1.5:1 | None | None | None | None |
| 3:1 | 2,4,6,..., | 3,5,7,..., | 5,9,...,[b] | - - |
| 5:1 | 2,3,4,5,...,[a] | 2,3,4,5,..., | 2,3,4,5,..., | 2,3,4,5,..., |
| 6:1 | All | All | All | All |

[a] Figure 4-26a.

[b] Figure 4-26b.

   It is evident from Figs. 4-25 and 4-26 that marginal passage
of a pulse through a widening leads to time delays of the order
0. 5-1 msec.   This would seem significant for the processing of
auditory information.   These delays can be induced by "varicosities" as indicated in Fig. 4-23b, and Bogoslovskaya, Lyubinskii,
Pozin, et al. (1973) have shown that such varicosities are clearly
evident in the dendrites of cochlear neurons of certain animals
(see Fig. 4-27).   This leads one to suspect that the dendrites may
play an important role in the processing of pulse train information.
We return to this idea in Chapter 6.
   On the axonal side Revenko, Timin, and Khodorov (1973) have
recently investigated the propagation of an action potential from
a myelinated fiber into a nonmyelinated terminal section.   This is
a region of low safety factor because the exposed fraction of the
membrane, f in (4.5.2), jumps from a low value on the myelinated fiber to unity on the terminal section.   To ensure conduction,
a narrowing of the fiber diameter by a factor of about three is required; even in this case frequency reduction similar to that displayed in Fig. 4-26 should be expected.

## 7.   DECREMENTAL CONDUCTION

   We have seen in Fig. 4-4 that the effect of multiplying $\bar{g}_{Na}$
and $\bar{g}_K$ by a "narcotizing factor" $\eta < 1$ is to reduce the speed

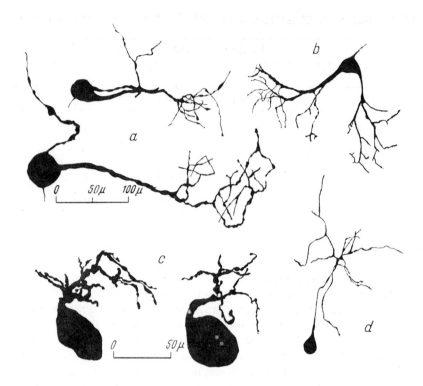

FIGURE 4-27.    Cochlear neurons of: (a) monkey, (b) hedgehog,
(c) owl, and (d) bat (Bogoslovskaya, Lyubinskii,
Pozin, et al. 1973).

and amplitude of the fast traveling-wave solution and to increase
the speed and amplitude of the slow threshold pulse.   At a critical
value of narcotization $\eta_C = 0.261$, these two pulse solutions of
the Hodgkin-Huxley equations merge.   For $\eta < \eta_C$, no traveling-
wave pulse solution has been found.   If the axon is suitably stimulated with $\eta < \eta_C$, a "decremental" pulse is observed that propagates with diminishing amplitude.   Figure 4-28, for example,
presents numerical calculations of decremental propagation on the
Hodgkin-Huxley axon with $\eta = 0.25$ by Cooley and Dodge (1966);
similar results have been obtained by Leibovic and Sabah (1969)
and by Khodorov, Timin, Vilenkin, et al. (1970) [see also the recent books by Leibovic (1972) and Khodorov (1974)].

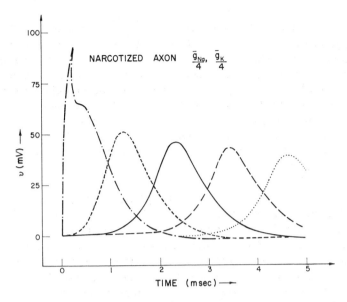

$v$ (mV) —

NARCOTIZED AXON   $\dfrac{\bar{g}_{Na}}{4}, \dfrac{\bar{g}_K}{4}$

TIME  (msec) —

FIGURE 4-28.   Propagation of a decremental pulse on a Hodgkin-
Huxley axon narcotized by a factor of 0.25.
Curves are voltage waveforms at 1 cm intervals
(redrawn from Cooley and Dodge, 1966).

These numerical studies are of considerable interest because
the possibility of decremental conduction was seriously challenged
a half century ago by Kato (1924).   His argument had a theoretical
foundation which he recently summarized as follows (Kato, 1970):

> The decrement theory and the decrementless theory can be
> said to rest on fundamentally different theoretical stand-
> points.   The former seems to hold that an impulse is con-
> ducted by its initial energy supplied <u>only</u> at its start which,
> being consumed as it propagates and having no supply of
> new energy during its conduction, gradually diminishes,
> in other words, suffers decrement . . . but the latter theory
> postulates that the impulse is conducted by means of new
> energy necessary for conduction produced <u>locally</u> as it is
> propagated, so it can be conducted in a decrementless
> manner.   What is important here is that, while those sup-
> porting the decrement theory held that the nerve (a living
> tissue in general) would change its nature qualitatively

when condition changed, we, in our decrementless theory, inferred that it would remain unchanged in its nature under changed conditions, and only be subject to quantitative change.    Therefore, our way of thinking was fundamentally different from theirs, not being concerned merely with the phenomenal fact as to whether conduction is decremental or decrementless (nondecremental).    In this sense the discussion on the right or wrong of our theory came to exert a great influence on the fundamental thought in the biological world.

To establish the validity of his perception, Kato prepared an experimental demonstration for the 1926 International Physiological Congress in Stockholm.    Using sciatic nerve muscle preparations (see Fig. 1-1) from Dutch frogs (his Japanese toads had perished during the long trip across Siberia), he first showed that the extinction times under narcotization were almost exactly equal for 1. 5 cm and 3. 0 cm of nerve.    Next he demonstrated a threshold effect for stimulation in the narcotized region.    Finally came the crucial "cut experiment" designed to demonstrate that such graded responses as were evident from stimulation in the narcotized region were due to electrical leakage rather than decremental conduction.    It was a terminal experiment since the narcotized region was to be cut precisely at the point of electrical stimulation in order to show that mechanical stimulation would not produce the same effect.    Thomas Kuhn (1962) has not described a more dramatic example of the establishment of a scientific paradigm than that moment so vividly recalled by Kato (1970).

> When Dr. Uchimura, just after applying the electrical stimulation, took a cutting pose, there came a voice from far behind of the room, "No muscle state can be seen from here! " Indeed, many of the observers wanted to witness whether the muscle would contract or not with their own eyes. It was Dr. Buytendijk, Professor of the University of Groningen, Holland, who offered to announce if the muscle moved or not, because he was nearest to the table.    Dr. Uchimura took up scissors once again and brought them near the nerve to cut it.    But his hand was trembling; that might cause some sort of straining on the nerve to make the muscle contract.    I had no courage to see the cutting instant.    Seconds fled.    Suddenly sounded,

"Keine Zuckung!! No twitch!! " It was Prof. Buytendijk's voice. And then followed another voice, "Revolution der Physiologie! ", from whom I could not identify. Scholar after scholar they presented me with congratulations and handshakes.

Thirty-three years later Lorente de Nó and Condouris (1959) were to lament the premature demise of the decremental conduction doctrine in the face of these objections.

It seems unbelievable, but it is true, that although the objections were soon proved to be unjustified by a number of authors [refs. ] , the doctrine of decremental conduction disappeared from the literature and there remained, to mold the thinking of neurophysiologists, only the all-or-nothing law in the generalized, inflexible form given to it by Kato . . . .

The theoretical objections posed by Kato to the qualitatively different aspect of decremental conduction disappear when one considers nonlinear pulse propagation on the nerve fiber to be dominated by the <u>power-balance condition</u>

$$P = uE \qquad\qquad (4.7.1)$$

where $P$ is the power (J/sec) consumed by the pulse, $E$ is the stored energy (J/m) released by the pulse, and $u$ is the pulse velocity (m/sec). The curves in Figs. 4-4 and 4-9 indicate those values of the plotted parameters for which a traveling-wave pulse satisfying (4.7.1) does exist. Since the upper branch is stable and the lower branch is unstable it seems reasonable to speculate that the inside $\oplus$ region is where pulse solutions can be found with $uE > P$. Conversely, in the outside $\ominus$ region we expect $uE < P$ for all pulse solutions. For $\eta$ slightly less than $\eta_C$, we expect to find a decremental pulse for which $uE$ is slightly less than $P$. As an example, take the FitzHugh-Nagumo equation (4.3.1) with $a = 0$ and $b = 0$. Putting "conservative" terms on the left and "dissipative" terms on the right, this can be written

$$-V_{tt} - \epsilon V = F'(V)V_t - V_{xxt} \qquad\qquad (4.7.2)$$

The conservative terms are recognized because they can be derived by substituting the Lagrangian density $\frac{1}{2}(\epsilon V^2 - V^2 t)$ into the Euler equation [Scott (1970)]. Then the corresponding energy density is $\frac{1}{2}(\epsilon V^2 + V_t^2)$ so the total energy is

$$\mathcal{E} = \frac{1}{2} \int_{-\infty}^{\infty} (\epsilon V^2 + V_t^2) dx \qquad (4.7.3)$$

Differentiating (4.7.3) with time and substituting (4.7.2) gives

$$\frac{\partial \mathcal{E}}{\partial t} = - \int_{-\infty}^{\infty} (V_{xt}^2 + F'(V)V_t^2) dx \qquad (4.7.4)$$

Along the traveling-wave locus in Fig. 4-9, the right-hand side integral must be zero. In the $\oplus$ region, the right-hand side is negative and in the $\ominus$ region it is positive. Perturbative techniques would seem to be appropriate for estimating the right-hand side near the traveling-wave locus. Some problems for which such a perturbative approach might be useful are listed below, and the corresponding critical parameter is indicated.

1.  For the FitzHugh-Nagumo equation, the critical parameter might be $\epsilon$ (as in Fig. 4-9) or a perturbation of $F(v)$ that reduces the negative area ($A_2$ in Fig. 4-6).

2.  For the Markin-Chizmadzhev model (see Fig. 4-16) the critical parameter might be $V_1$, or from (4.4.7) $r_s$ or $c$.

3.  The amplitude of the Markin-Chizmadzhev pulse as determined by (say) $J_1$ in (4.4.7) might be chosen as the critical parameter. Then the description would apply to the decrement for pulses below the threshold level.

4.  The parameter $\alpha$ for the representation of myelinated propagation indicated in Fig. 4-21.

5.  Various real experimental parameters, such as temperature, ionic concentrations, and narcotic concentrations, which would modify the constants in the Hodgkin-Huxley expression (4.1.2) for ionic current.

6.  The wavelength, $\lambda$, for periodic traveling-wave solutions such as those by Rinzel and Keller (1973) shown in Fig. 4-14.

In view of all these possibilities for application, the development of a formal perturbation theory for decremental conduction is of great current interest.   To introduce the fundamental ideas it is convenient to return to the K. P. P. equation

$$v_{xx} - v_t - F(v) = 0 \qquad (4.7.5)$$

and add a small perturbation, $\delta$, so it becomes

$$v_{xx} - v_t - F(v) = \delta \qquad (4.7.6)$$

Following the notation of Whitham (1974) (pp. 493-497) for "two timing" and "double crossing" we write

$$v = V_0(\theta, X, T) + \delta V_1(\theta, X, T) + \cdots \qquad (4.7.7)$$

where $X \equiv \delta x$ and $T \equiv \delta t$ are slow space and time variables.   It is assumed that $v$ depends on fast space and time ($x$ and $t$) through

$$\theta = \delta^{-1} \Theta(X, T) \qquad (4.7.8)$$

and for notational convenience we define

$$k(X, T) \equiv \Theta_X \quad \text{and} \quad \nu(X, T) \equiv \Theta_T \qquad (4.7.9)$$

Then to first order in $\delta$

$$v_{xx} = k^2 V_{0,\theta\theta} + \delta(k^2 V_{1,\theta\theta} + 2k V_{0,\theta X} + k_X V_{0,\theta})$$

$$v_t = \nu V_{0,\theta} + \delta(\nu V_{1,\theta} + V_{0,T}) .$$

Thus equating coefficients in (4.7.6) for the first two orders in $\delta$ gives

$$\delta^0 : k^2 V_{0,\theta\theta} - \nu V_{0,\theta} - F(V_0) = 0 \qquad (4.7.10)$$

$$\delta^1 : k^2 V_{1,\theta\theta} - \nu V_{1,\theta} - F'(V_0)V_1 = 1 + V_{0,T} - 2k V_{0,\theta X} - k_X V_{0,\theta}$$

$$(4.7.11)$$

Equation (4. 7. 11) is a linear, nonhomogeneous ordinary differential equation for $V_1$. It must have a solution for the perturbation expansion (4. 7. 7) to be valid. To investigate this requirement, suppose it is written in the abbreviated form

$$LV_1 = f \qquad (4.7.11')$$

Here L is the differential operator

$$L \equiv k^2 \frac{d^2}{d\theta^2} - \nu \frac{d}{d\theta} - F'(V_0) \qquad (4.7.12)$$

and f is the inhomogeneous term that can be calculated from $V_0$. Using the notation for an "inner product" of two functions [ say $v(\theta)$ and $w(\theta)$]

$$(v, w) \equiv \int_{-\infty}^{\infty} v(\theta)w(\theta)d\theta \qquad (4.7.13)$$

recall that (if v and w $\rightarrow 0$ as $\theta \rightarrow \pm\infty$) the "adjoint" $(\tilde{L})$ of an operator (L) is defined by the requirement

$$(v, Lw) = (\tilde{L}v, w) \qquad (4.7.14)$$

Suppose next that $\tilde{\phi}$ is a solution of the adjoint homogeneous equation

$$\tilde{L}\tilde{\phi} = 0 \qquad (4.7.15)$$

which satisfies the condition $\tilde{\phi} \rightarrow 0$ as $\theta \rightarrow \pm\infty$. Then <u>the inner product</u> $(\tilde{\phi}, f)$ <u>must equal zero for</u> (4. 7. 11) <u>to have a solution.</u> To see this, suppose $(\tilde{\phi}, f) \neq 0$. Then from (4. 7. 11')

$$(\tilde{\phi}, LV_1) = (\tilde{\phi}, f) \neq 0$$

$$= (\tilde{L}\tilde{\phi}, V_1) = 0$$

which is a contradiction.

Thus in order for (4. 7. 11) to have a solution, the zero-order solution, $V_0$, must evolve in such a way that the condition

$$(\widetilde{\phi}, f) = 0 \qquad (4.7.16)$$

is satisfied.  This condition can be used to determine the behavior of $V_0$.

The first obstacle in using this result is to find an appropriate solution for the adjoint homogeneous equation (4.7.15).  To effect this, note that if $\phi$ is a solution of the homogeneous equation $L\phi = 0$, one can use integration by parts to show

$$\int \widetilde{\phi} \, L\phi \, d\theta - \int \phi \, \widetilde{L}\widetilde{\phi} \, d\theta = k^2(\widetilde{\phi}\phi' - \widetilde{\phi}'\phi) - \nu\widetilde{\phi}\phi \qquad (4.7.17)$$

Since the left-hand side is zero, we find $\widetilde{\phi} = \phi \exp(-\frac{\nu}{k^2}\theta)$.  Thus, if we know a solution of $L\phi = 0$, we can immediately write a solution for $\widetilde{L}\widetilde{\phi} = 0$.  A solution for $L\phi = 0$ is readily obtained by noting that differentiation of (4.7.10) gives

$$k^2(V_{0,\theta})_{\theta\theta} - \nu(V_{0,\theta})_\theta - F'(V_0)V_{0,\theta} = 0 \qquad (4.7.18)$$

Thus $\phi = V_{0,\theta}$ is a solution for $L\phi = 0$ and the corresponding solution of the adjoint equation is

$$\widetilde{\phi} = V_{0,\theta} \exp(-\frac{\nu}{k^2}\theta) \qquad (4.7.19)$$

Since the adjoint homogeneous equation is second order, it has two independent solutions.  From the asymptotic form of $\widetilde{L}$ as $\theta \to \pm\infty$, however, it is seen that the other solution does not approach zero as $\theta \to \pm\infty$.  Thus there is only a single secularity condition implies by (4.7.15).  It is

$$\int_{-\infty}^{\infty} V_{0,\theta} \exp(\frac{-\nu}{k^2}\theta)[1 + V_{0,T} - 2kV_{0,\theta X} - k_X V_{0,\theta}] \, d\theta = 0 \qquad (4.7.20)$$

To see that this condition implies for the functional behavior of $V_0$, consider a specific example.  Let

$$F(V) = V(V-a)(V-1) \qquad (4.7.21)$$

Then using the techniques of Section 4-2 [in particular, (4.2.11) and (4.2.12)], it is readily shown that (4.7.10) has the solution

$$V_0 = \frac{1}{1 + \exp(\theta/\sqrt{2}\,k)} \qquad (4.7.22)$$

with traveling-wave velocity

$$-\frac{\nu}{k} = \frac{1 - 2a}{\sqrt{2}} \qquad (4.7.23)$$

If $\delta = 0$, this is an exact traveling-wave solution of (4.7.6). If $\delta \neq 0$, then $\nu$ and $k$ vary with X and T. How do they vary? The secularity condition (4.7.22) must be satisfied and also the requirement

$$\nu_X = k_T \qquad (4.7.24)$$

for integrability of $\Theta$. We can assume that $k$ is independent of T and $\nu$ a constant (say $-1$) without violating (4.7.24). Since $V_0$ has explicit dependence only on $\theta$ and $k$, (4.7.20) and (4.7.24) can be solved for

$$k_X = \frac{\int_{-\infty}^{\infty} V_{0,\theta}\, \exp\left(\frac{-\nu}{k^2}\,\theta\right) d\theta}{\int_{-\infty}^{\infty} V_{0,\theta}\, \exp\left(-\frac{\nu}{k^2}\theta\right)[V_{0,\theta} + 2k\,V_{0,\theta,k}]\,d\theta} \qquad (4.7.25)$$

Then $V_0$ evolves according to (4.7.22) with a velocity $u(X) = k^{-1}$, where $k$ is obtained from integration of (4.2.25). In the "unstretched" distance scale

$$k_x = \delta k_X \qquad (4.7.26)$$

Thus $V_0$ speeds up or slows down if $\delta$ is increased from zero in one direction or the other.

The recent numerical studies of decremental conductance by Khodorov and his coworkers have been much more detailed. Their approach is to use the Hodgkin-Huxley equations in order to relate pulse propagation through a decremental region with biochemical manipulation. For example, Khodorov, Timin, Vilenkin, et al. (1970) have modeled an increased concentration of $Ca^{++}$ by a shift in voltage for $\alpha_n$, $\beta_n$, $\alpha_m$, and $\beta_m$ (see 3.25), as is implied by the measurements of Frankenhaeuser and Hodgkin (1957). With a 7 mV shift toward depolarization, the traveling-wave amplitude is reduced to 82 mV, but a 15 mV shift toward depolarization results in decremental conduction. Khodorov and Timin (1970, 1971) have modeled the effects of such influences as cooling, tetrodotoxin, and narcotics, on the propagation of pulse trains over segments of decremental paths. In general they observe a variety of rhythmical variations qualitatively similar to those displayed in Fig. 4-26 for widening tapers. This appears to be a most fruitful research area for future collaboration between biochemists and electrophysiologists.

## 8.  A SUPERCONDUCTIVE NEURISTOR

Before concluding our study of pulse propagation on a nerve fiber it may be appropriate to consider briefly an electronic analog: the neuristor. Coined by Crane (1962), this generic term implies a class of nonlinear transmission lines that share three critical properites of the active nerve fiber: (a) attenuationless propagation of a traveling-wave pulse, (b) threshold for excitation of the pulse, and (c) mutual pulse destruction on collision. The particular realization to be discussed here is a superconductive neuristor (Scott, 1964b; Parmentier, 1969). Pulse propagation on this system differs in an interesting way from that found from the Hodgkin-Huxley or FitzHugh-Nagumo models of a nerve fiber.
    Figure 4-29 is a photograph of the currently available device (Reible and Scott, 1975), which consists of a layer of sputtered niobium film overlaid with a narrow strip of tin (see Fig. 4-30a). These two layers are separated by a thin ($\sim 50$ Å) insulating barrier of niobium oxide through which electrons can pass via quantum-mechanical tunneling (Giaever and Megerle, 1962). The corresponding TLEC is sketched in Fig. 4-30b and, briefly, the elements of the system are as follows:

c – the capacitance of the insulating barrier per unit length in the direction of propagation;

j(v) – the nonlinear conduction current through the barrier per unit length;

r – the resistance per unit length for normal electron current flow parallel to the barrier;

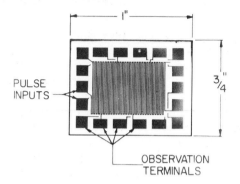

FIGURE 4-29. Superconductive neuristor (length 35 cm) on a glass substrate consisting of a tin strip (width 0.0064 cm) over a niobium line (width 0.048 cm) with an insulating barrier of niobium oxide.

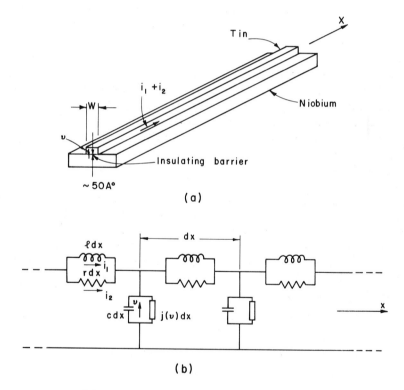

FIGURE 4-30. (a) Sketch of the superconductive neuristor, and (b) equivalent circuit.

144

$\ell$ - the inductance per unit length for superconducting current flow parallel to the barrier;

v - voltage across the barrier;

$i_1$ - superconducting current parallel to the barrier;

$i_2$ - normal current parallel to the barrier.

The total nonlinear tunnel current through the insulating barrier, $I_T(v)$, varies with voltage as shown in Fig. 4-31, but an adjustable bias current, $I_B$, can be introduced through appropriate terminals so that

$$j(v) = \frac{I_T(v) - I_B}{W} \qquad (4.8.1)$$

If the bias current is sufficiently large, $j(v)$ becomes zero at three values of voltage, as in Fig. 4-8a, and energy can be released in support of a traveling-wave pulse.

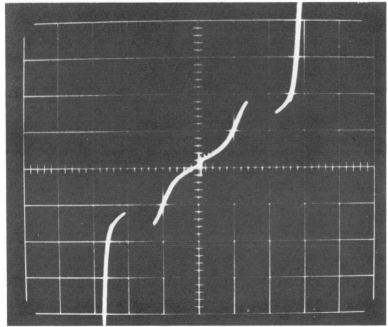

FIGURE 4-31.    Plot of total Giaever type tunneling current against voltage for a 35 cm tin-niobium neuristor at 2.7°K. Vertical scale: 20 mA/div; Horizontal scale: 0.5 mV/div.

The TLEC in Fig. 4-30b implies the PDEs

$$\frac{\partial v}{\partial x} = - ri_2$$

$$\frac{\partial i_1}{\partial x} + \frac{\partial i_2}{\partial x} = - c\frac{\partial v}{\partial t} - j(v) \qquad (4.8.2a, b, c)$$

$$\frac{\partial i_1}{\partial t} = \frac{r}{\ell} i_2$$

which are rather closely related to the FitzHugh-Nagumo equations (4.3.1). Assuming the variables $v, i_1$, and $i_2$ to depend only on the moving spatial coordinate $\xi = x - ut$, reduces (4.8.2) to the ODEs

$$\frac{dv}{d\xi} = - ri_2$$

$$\frac{di_2}{d\xi} = - rcu\left(1 - \frac{1}{\ell cu^2}\right) i_2 - j(v) \qquad (4.8.3a, b)$$

In the limit

$$\frac{1}{\ell} \to 0$$

(4.8.2) and (4.8.3) reduce to the corresponding PDEs and ODEs for the simple nonlinear diffusion (K. P. P.) equation given in (4.2.3) and (4.2.4). Thus in the case

$$\frac{1}{\ell} > 0$$

the traveling-wave solutions to (4.8.3) will correspond directly to those given by (4.2.4) and illustrated in Figs. 4-8, 4-9, and 4-15. If $u_a$ is the velocity of a certain solution of (4.2.4), then the corresponding solution of (4.8.3) will have a velocity, $u$, given by

$$u_a = u\left(1 - \frac{1}{\ell cu^2}\right)$$

or

$$u = \tfrac{1}{2}\left[u_a \pm \sqrt{u_a^2 + 4/\ell c}\,\right] \qquad (4.8.4)$$

For typical superconductive neuristors, $1/\sqrt{\ell c}$ is measured to be about 1/20 the velocity of light (Yuan and Scott, 1966) while $u_a$ is the order of magnitude of the pulse velocity on a nerve fiber. Thus the inequality

$$\frac{1}{\sqrt{\ell c}} \gg u_a \qquad (4.8.5)$$

is satisfied by several orders of magnitude and (4.8.4) reduces to

$$u \approx \frac{1}{\sqrt{\ell c}} + \tfrac{1}{2}u_a \qquad (4.8.6)$$

Referring back to Figs. 4-6 and 4-7, appropriate values for $u_a$ are $\pm u_1$ and $\pm u_2$ for the propagation of level changes and $0$ for the pulse. From (4.8.1), however, the ratio of areas $A_1$ and $A_2$ can be changed by a simple adjustment of the bias current. For convenience we can define $I_c$ as the critical value of bias current at which $A_1 = A_2$. Then

$$I_B > I_c \implies A_2 > A_1$$

$$I_B < I_c \implies A_2 < A_1$$

as indicated on Fig. 4-32. This figure shows the velocity, from (4.8.6), plotted against the bias current for the traveling-wave solutions on the neuristor, which correspond to those in Figs. 4-6 and 4-7.

The pulse solution is shown in Section 5-1 to be unstable (Lindgren and Buratti, 1969). Measurements by Reible on an eighty-two-section electronic analog of (4.8.2) show that such a pulse either decays to zero or grows to a metastable pulse in which the leading edge travels slightly faster than the trailing edge (see Fig. 4-33). From such data the relative velocities of the leading and trailing edges can be readily measured and plotted as a function of bias current. As shown in Fig. 4-34, such measurements confirm (4.8.6) and Fig. 4-32.

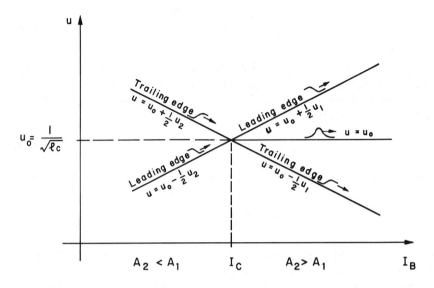

FIGURE 4-32.    Plot of velocity (u) against bias current ($I_B$) for
traveling-wave solutions of (4.8.2).

Thus the dynamic behavior of (4.8.2) for the superconducting
neuristor presents an interesting contrast to that of the FitzHugh-
Nagumo equation.    The "center of mass" for the metastable pulse
has the same velocity as that of the unstable pulse.    It absorbs
the extra energy necessary to satisfy the power-balance condition
(4.3), P = uE, not by going faster, but by growing fatter!

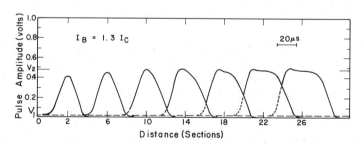

FIGURE 4-33.    Evolution of a pulse waveform for a bias level of
$I_B = 1.3 I_C$ measured on an eighty-two-section
electronic analog of (4.8.2).

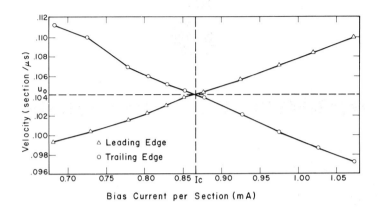

FIGURE 4-34.   Leading- and trailing-edge velocities for pulses
              on the eighty-two-section analog plotted against
              current (data confirms the predictions of Fig. 4-33).

149

# 5

# Stability and Threshold

Companion of my griefs! thy sinking frame
Had often drooped, and then erect again
With shews of health had mocked forebodings dark;
Watching the changes of that quivering spark,
I feared and hoped, and dared to trust at length,
Thy very weakness was my tower of strength.

Mary Wollstonecraft Shelley

In Chapter 4 we sought traveling wave solutions for various
types of nonlinear diffusion equations, and we expect that such
solutions will help us to understand the "level 3" relation between
neuron behavior and membrane electrodynamics. We found stationary solutions in the form of both solitary and periodic waves. Our
analytic technique was to assume that dependent variables are
functions of $x$ and $t$ only through the argument $\xi = x - ut$ as
indicated in (4.1.3). This is equivalent to introducing the independent variable transformation (4.1.4) and then assuming no dependence on $\tau$ (i.e. $\partial/\partial\tau = 0$). Having found such traveling-
wave solutions, it is important to know whether or not they are
stable with respect to perturbations that might reasonably be expected to arise in an experimental situation. To study the time
evolution of such perturbations, it is necessary to consider the $\tau$
dependence.

However, it should not be assumed that unstable traveling-
wave solutions are uninteresting. An unstable solitary wave, for
example, can be expected either to grow into a stable solitary
wave or collapse to zero under the influence of small perturbation.
Since it thus represents a "watershed" or "divide" from which the
system can flow toward one of two stable states; the unstable

150

solution determines the threshold conditions necessary to induce a stable solution.

## 1.  WAVEFORM STABILITY

To introduce the basic ideas of waveform stability analysis, we investigate the K. P. P. form of the nonlinear diffusion equation (4.1),

$$V_{xx} - V_t = F(V) \qquad (5.1.1)$$

Traveling-wave solutions for (5.1.1) were considered in detail in Section 4-2 assuming a cubic form for the function $F(V)$. Equation (5.1.1) is sufficiently simple for exposition and the results to be obtained serve as a basis for stability investigation of the Fitz-Hugh–Nagumo and Hodgkin–Huxley traveling waves.

Under the transformation (4.1.4), (5.1.1) becomes

$$V_{\xi\xi} + uV_\xi - V_\tau = F(V) \qquad (5.1.2)$$

where $V$ is now considered a function of $\xi$ (space in a coordinate system moving with velocity $u$) and $\tau$ (the same time scale as $t$). The traveling-wave solution $V_T(\xi)$ must satisfy

$$V_{T,\xi\xi} + uV_{T,\xi} = F(V_T) \qquad (5.1.3)$$

and a general solution of (5.1.2) can be considered as the sum of a traveling-wave solution and a perturbation $V_P(\xi,\tau)$. Thus

$$V(\xi,\tau) = V_T(\xi) + V_P(\xi,\tau) \qquad (5.1.4)$$

Substituting (5.1.4) into (5.1.2) gives

$$V_{P,\xi\xi} + uV_{P,\xi} - V_{P,\tau} = F(V_p + V_T) - F(V_T) \qquad (5.1.5)$$

as a nonlinear and $\xi$ dependent PDE for the evolution of the perturbation. It is important to recognize that no approximations have been made in going from (5.1.1) to (5.1.5).

Investigation of (5.1.5) for the evolution of $V_P(\xi,\tau)$ subject

to prescribed initial and boundary conditions constitutes the
"waveform stability problem" for a traveling-wave solution of
(5. 1. 2) with velocity u. This equation has been studied in connection with the propagation of: (a) flames (Zeldovich and Barenblatt, 1959; Kanel', 1962), (b) "Gunn effect" domains in bulk
semiconductors (Knight and Peterson, 1967; Eleonskii, 1968), and
(c) traveling waves on "neuristors" and electronic analogs for the
nerve fiber (Parmentier, 1967, 1968, 1969, 1970; Buratti and Lindgren, 1968; Lindgren and Buratti, 1969; Maginu, 1971).

One approach to the study (5. 1. 5) is to assume the perturbation small enough so that the right-hand side can be approximated
by

$$F(V_P + V_T) - F(V_T) \approx \left.\frac{dF}{dV}\right|_{V = V_T} \times V_P \equiv G(V_T)V_P \qquad (5.1.6)$$

whereupon (5. 1. 5) is "linearized" to

$$V_{P, \xi\xi} + uV_{P, \xi} - V_{P, \tau} = G[V_T(\xi)]V_P \qquad (5.1.7)$$

Elementary solutions to (5. 1. 7) will either decay exponentially
with time, grow exponentially with time, or remain constant.
Thus, with respect to the linearized equation, we can say the
system is: (a) asymptotically stable if all elementary solutions
decay, (b) unstable if any elementary solution grows, and
(c) stable if both (a) and (b) are not satisfied.

This is a neat scheme but we must be wary of drawing conclusions from (5. 1. 7) that are not relevant to the application of
(5. 1. 5) in a real situation. While we might conclude asymptotic
stability with respect to (5. 1. 7), for example, it may not be reasonable to assume perturbations sufficiently small for (5. 1. 7) to
apply. As Eckhaus (1965) puts it, "infinitesimal disturbances are
certainly unavoidable, but not all unavoidable disturbances may
be considered infinitesimal. "[*] On the other hand, if (5. 1. 7) indicates elementary solutions that grow, these may eventually be
bounded by the nonlinear character of (5. 1. 5). Such a bound may
be so close to the original solution that the system is, in a practical sense, stable. With these caveats in mind, let us proceed
to the analysis of (5. 1. 7).

---

[*] Those who experiment with real nerve fibers will probably agree.

If $V_P$ is constructed from elementary product solutions of the form

$$V_P \sim \phi(\xi) e^{-sT} \tag{5.1.8}$$

then $\phi$ must satisfy the eigenvalue equation

$$\phi_{\xi\xi} + u\phi_\xi + \{s - G[V_T(\xi)]\}\phi = 0 \tag{5.1.9}$$

The condition for asymptotic stability is that all the eigenvalues, $s$, which are allowed for solutions (5.1.9) must have positive real parts. This would require the magnitude of the corresponding elementary solution (5.1.8) to decay exponentially with time. In a certain sense asymptotic stability is never possible. This can be seen by differentiating (5.1.3) for the traveling-wave solution with respect to $\xi$ to obtain

$$(V_{T,\xi})_{\xi\xi} + u(V_{T,\xi})_\xi - G(V_T)V_{T,\xi} = 0 \tag{5.1.10}$$

and noting that this is the same equation obeyed by $\phi$ when $s = 0$. Thus the eigenfunction of (5.1.9) with zero eigenvalue is

$$\phi = V_{T,\xi} \quad \text{for} \quad s = 0 \tag{5.1.11}$$

The physical meaning of this result is seen by considering an infinitesimal translation, $\alpha$, of $V_T$ along the $\xi$-axis. Since

$$V_T(\xi + \alpha) = V_T(\xi) + \alpha V_{T,\xi} \tag{5.1.12}$$

this is equivalent to adding an infinitesimal amount of the $s = 0$ eigenfunction. But we expect a translation perturbation to neither grow nor decay. The observation that the perturbation eigenfunction corresponding to zero eigenvalue is the derivative of the traveling wave is quite general and not at all restricted to solutions of (5.1.1). Many investigators avoid this situation by defining stability with respect to a metric that permits arbitrary translations with $\xi$ (Zeldovich, and Barenblatt, 1959; Kanel', 1962; Maginu, 1971; Evans, 1972; Sattinger, 1976).

Next it is of interest to determine whether $s = 0$ is the lowest eigenvalue; if it is not, (5.1.8) indicates instability. We make this determination with respect to the boundary conditions

$$\phi \to 0 \quad \text{as} \quad |\xi| \to \infty \tag{5.1.13}$$

If the change of dependent variable

$$\phi = [\exp - (\tfrac{u}{2})\xi]\psi \tag{5.1.14}$$

is introduced into (5. 1. 9) (parmentier, 1967), $\psi$ must satisfy the Schrödinger equation

$$\psi_{\xi\xi} + \{s - \frac{u^2}{4} - G[V_T(\xi)]\}\psi = 0 \tag{5.1.15}$$

for which the eigenvalues are real and bounded from below [Morse and Feshbach (1953) pp. 766-8]. If $s \leq 0$, $G \to G_1 > 0$ as $\xi \to +\infty$, and $G \to G_2 > 0$ as $\xi \to -\infty$, $\psi$ must also satisfy the boundary condition (5. 1. 13). Then $s = 0$ is the lowest eigenvalue if the corresponding eigenfunction, $(dV_T/d\xi)$, has no zero crossings. This condition is satisfied for the "level change" waves in Fig. 4-6 but not for the pulse wave in Fig. 4-7. Thus the smooth level change waves are stable with respect to the linearized equation, but any solution for which $V_T$ is not monotone increasing or decreasing with $\xi$ will have eigenvalues $s < 0$, and, from (5. 1. 8), will be unstable. This conclusion is independent of the form of the function $F(V)$ in (5. 1. 1).

This result has been extended to perturbations that are not infinitesimal by Maginu (1971). He expresses the right-hand side of (5. 1. 5) by a Taylor series so that

$$V_{P,\xi\xi} + uV_{P,\xi} - V_{P,\tau} = F'(V_T)V_P + \tfrac{1}{2}F''(V_T)V_P^2 + \cdots \tag{5.1.16}$$

for $V_P$ within the appropriate range of convergence. Then he finds a set of functions

$$\{V_P^{(n)}\}$$

with the property that as $\tau \to \infty$

$$V_P^{(n)} \to \frac{\alpha^n}{n!} \frac{d^n V_T}{d\xi^n}$$

This asymptotic property is established by requiring the partial sum

$$V_P^{(1)} + V_P^{(2)} + \cdots + V_P^{(n)}$$

to satisfy (5.1.16) when the Taylor series is approximated up to terms of order $\alpha^n$. Then, taking $V_P$ as the infinite sum

$$V_P = \sum_{n=1}^{\infty} V_P^{(n)} \tag{5.1.17}$$

it is seen that

$$V_P \rightarrow V_T(\xi + \alpha) - V_T(\xi) \quad \text{as} \quad \tau \rightarrow \infty \tag{5.1.18}$$

This is asymptotic nonlinear stability with respect to a metric that permits translations in the $\xi$-direction. The only restriction on $V_P$ is that it must lie within the range of convergence in (5.1.16).

To see how this argument goes, note first that we have already demonstrated [through analysis of (5.1.7)], that $V_P^{(1)} \rightarrow \alpha V_{T,\,\xi}$ for $\tau \rightarrow \infty$ as long as $V_T$ is a monotone increasing or decreasing function of $\xi$.

To second order, $V_P^{(2)}$ must satisfy

$$V_{P,\,\xi\xi}^{(2)} + uV_{P,\,\xi}^{(2)} - V_{P,\,\tau}^{(2)} = F'(V_T)V_P^{(2)} + \tfrac{1}{2}F''(V_T)(V_P^{(1)})^2 \tag{5.1.19}$$

Differentiating (5.1.3) twice with respect to $\xi$ gives

$$V_{T,\,\xi\xi\xi\xi} + uV_{T,\,\xi\xi\xi} = F'(V_T)V_{T,\,\xi\xi} + F''(V_T)V_{T,\,\xi}^2 \tag{5.1.20}$$

The variable

$$w \equiv V_P^{(2)} - \tfrac{1}{2}\alpha^2 V_{T,\,\xi\xi} \tag{5.1.21}$$

obeys the equation [ (5.1.19) $- \tfrac{1}{2}\alpha^2$ (5.1.20)]   or

$$w_{\xi\xi} + uw_\xi - w_\tau = F'(V_T)w + \tfrac{1}{2}F''(V_T)[(V_P^{(1)})^2 - \alpha^2 V_{T,\,\xi}^2] \tag{5.1.22}$$

But, as $\tau \to \infty$, this approaches

$$w_{\xi\xi} + uw_\xi - w_\xi = F'(V_T)w \tag{5.1.23}$$

which is identical to (5.1.7) so $w \to \alpha_1 V_{T,\xi}$.
    Then from (5.1.21)

$$V_P^{(1)} + V_P^{(2)} \to (\alpha + \alpha_1)V_{T,\xi} + \tfrac{1}{2}\alpha^2 V_{T,\xi\xi} \tag{5.1.24}$$

as $\tau \to \infty$. The addition of $\alpha_1$ to $\alpha$ in the first term constitutes a second-order correction to the translation caused by the initial perturbation. It can be absorbed by redefining $\alpha$ as $(\alpha + \alpha_1)$ in (5.1.21) and (5.1.22). Higher-order estimates are treated in a similar manner.
    Consider finally the nonlinear bounds on those traveling waves, $V_T(\xi)$, that are not monotone increasing and hence unstable with respect to the linearized equation (5.1.7). These will grow no further than the stable, monotone increasing transition wave and will decay no further than zero. It seems reasonable to suppose that these are the bounds of interest.
    It should be emphasized that these conclusions do not apply to transition waves between 0 and $V_1$ in Fig. 4-7. Since the singular point at $V_1$ corresponds to negative differential conductance of the membrane, it is unstable even under space-clamped conditions. The stability of such waves is studied in connection with a problem of genetic diffusion where the dependent variable must be less than or equal to its value at the singular point (Fisher, 1937; Kolmogoroff, Petrovsky, and Piscounoff, 1937; Canosa, 1973; Rosen, 1974). Aronson and Weinberger (1975) have compared the asymptotic behavior of (5.1.1) for $F(V)$ equal to $V(1 - V)$ with that for $F(V)$ equal to $V(1 - V)(V - V_1)$.
    Lindgren and Buratti (1969) have investigated the stability of traveling-waves on an exponentially tapered version of the non-linear diffusion equation. From (4.6.3a)

$$V_{xx} - \gamma V_x - V_t = F(V) \tag{5.1.25}$$

where $\gamma$ is the tapering exponent indicated in (4.6.1). Traveling-wave solutions identical to those indicated in Figs. 4-6, 4-7, and 4-13 are readily obtained. The only difference is that the traveling-wave velocity is

$$u = u_0 + \gamma \qquad (5.1.26)$$

where $u_0$ is the velocity that would be calculated for no tapering as indicated in (4.6.4). Expressing a perturbation of a traveling-wave as in (5.1.5), the corresponding PDE for its evolution is

$$V_{P,\xi\xi} + (u - \gamma)V_{P,\xi} - V_{P,\tau} = F(V_P + V_T) - F(V_T) \qquad (5.1.27)$$

Lindgren and Buratti express the right-hand side of (5.1.27) as

$$F(V_P + V_T) - F(V_T) = G(V_T)V_P + \epsilon\,(\xi,\,V_P) \qquad (5.1.28)$$

where the "remnant" $\epsilon\,(\xi, V_P)$ has the property

$$\lim_{V_P \to 0} \frac{|\epsilon\,(\xi, V_P)|}{|V_P|} = 0 \qquad (5.1.29)$$

Then they define the Liapunov functional (Hahn, 1963)

$$L(V_P) = \tfrac{1}{2} \int_{-\infty}^{\infty} W^2(\xi)V_P^2(\xi, \tau)d\xi \qquad (5.1.30)$$

where $W(\xi)$ is an appropriate weighting function to be determined. Assuming the initial perturbation, $V_P(\xi, 0)$, to have finite energy, stability is ensured if $\dot{L} = (dL/dt) \leq 0$ for $\tau > 0$. Asymptotic stability requires $\dot{L} < 0$.

Differentiating (5.1.30) with respect to time and substituting (5.1.27) yields

$$\dot{L} = \int_{-\infty}^{\infty} W^2 V_P[V_{P,\xi\xi} + (u - \gamma)V_{P,\xi} - G(V_T)V_P - \epsilon\,(\xi, V_P)]d\xi \qquad (5.1.31)$$

For a linear stability analysis, the remnant $\epsilon$ is neglected and, after integration by parts, (5.1.31) assumes the form

$$\dot{L} = -\int_{-\infty}^{\infty} \{\psi_\xi^2 + [(u-\gamma)\frac{W_\xi}{W} - (\frac{W_\xi}{W})^2 + G(V_T)] \psi^2\} d\xi \qquad (5.1.32)$$

where

$$\psi \equiv W V_P \qquad (5.1.33)$$

Now it is convenient to choose the weighting function $W(\xi)$ so $\dot{L}$ is minimized. Thus $(W_\xi/W) = \frac{1}{2}(u - \gamma)$ or

$$W = \exp[\frac{1}{2}(u - \gamma)\xi] \qquad (5.1.34)$$

and (5.1. 2) becomes

$$\dot{L} = -\int_{-\infty}^{\infty} \{\psi_\xi^2 + [\frac{1}{4}(u - \gamma)^2 + G(V_T)] \psi^2\} d\xi \qquad (5.1.35)$$

For stability we must ensure that $\dot{L}$ does not become positive. To investigate this possibility we can consider the variational problem of maximizing (5.1.35) subject to the energy constraint

$$\int_{-\infty}^{\infty} \psi^2 d\xi = \text{const} \qquad (5.1.36)$$

For this condition (Morse and Feshbach, 1953) $\psi$ must satisfy the Euler equation

$$\psi_{\xi\xi} + [s - \frac{1}{4}(u - \gamma)^2 - G(V_T)] \psi = 0 \qquad (5.1.37)$$

Integrating (5.1.35) by parts and substituting (5.1.37) gives the maximum value for $\dot{L}$ as

$$\dot{L} \leq -\int_{-\infty}^{\infty} s\psi^2 d\xi \qquad (5.1.38)$$

Thus if (5.1.37) has only positive eigenvalues, the linear stability analysis would imply absolute stability. The corresponding computation for nonlinear stability assumes $\epsilon(\xi, V_p) \neq 0$, where

$$\dot{L} \leq -\int_{-\infty}^{\infty} [s + \epsilon(\xi, V_p)/V_p]\psi^2 d\xi \tag{5.1.39}$$

From the property of the remnant as $V_p \to 0$ that was expressed in (5.1.29), it is clear that linear absolute stability $[\dot{L} < 0$ from (5.1.38)] implies nonlinear absolute stability $[\dot{L} < 0$ from (5.1.39)]. It is only necessary to choose $V_p$ sufficiently small enough so that $|\epsilon/V_p| < s$ for any eigenvalue of (5.1.37).

However, as we noted above in connection with (5.1.9), there is an eigenfunction of (5.1.38) with $s = 0$, and perturbation with this eigenfunction simply translates $V_T$ along the $\xi$-axis. But if we define $V_p$ as the difference between $V(\xi, \tau)$ and a translated traveling-wave $V_T \xi + \alpha$ where $\alpha$ is adjusted to minimize the difference, then the $s = 0$ eigenfunction disappears from the analysis. With this definition of $V_p$ the transition waves in Fig. 4-6 are stable, whereas the pulse wave in Fig. 4-7 and the periodic waves in Fig. 4-13 are unstable.

For the superconductive neuristor described in Section 4-8, Lindgren and Buratti (1969) choose the Liapunov functional as the sum of electric and magnetic energies or

$$L = \tfrac{1}{2}\int_{-\infty}^{\infty} (cv_p^2 + \ell i_{1p}^2) d\xi \tag{5.1.40}$$

where $v_p$ and $i_{1p}$ are perturbations of $v(\xi)$ and $i_1(\xi)$ which satisfy (4.8.3). Then

$$rc\dot{L} = -\int_{-\infty}^{\infty} \{v_{p,\xi}^2 + rG[v(\xi)]\, v_p^2\} d\xi \tag{5.1.41}$$

with $G[v(\xi)]$ defined as in (5.1.6). The form of (5.1.41) is identical to that of (5.1.35), so a corresponding stability argument can be developed. In particular the transition wave-forms on the sloping branches of Fig. 4-32 are stable, whereas the pulse on the horizontal branch is unstable.

A corresponding stability investigation for a traveling-wave

solution of the FitzHugh-Nagumo equation (4.3.1) is considerably
more difficult because the linearized problem is of third order.
Thus the eigenvalue problem, corresponding to (5.1.9), cannot be
made self-adjoint and the eigenvalues are generally complex. The
eigenfunction for $s = 0$ is still $V_{T, \xi}$, but there is no simple relation between the number of zero crossings of the eigenfunctions
and the order of the real parts of the corresponding eigenvalues.
However, we have already shown branches ① and ③ of the
singular orbit Ⓐ in Fig. 4-11a to be stable, which is consistent
with the numerical results of FitzHugh (1969) and Rinzel and Keller
(1973) indicating stability along the high-velocity branch for particular functions $F(V)$.

In a series of papers Evans (1972) has investigated a generalization, of the Hodgkin-Huxley equations with the form suggested
by FitzHugh (1969)

$$V_{xx} - V_t = F_0(V, w_1, \cdots, w_n) \qquad (5.1.42)$$

$$w_{i,t} = F_i(V, w_1, \cdots, w_n) \qquad i = 1, \cdots, n$$

where the $F$ values are twice continuously differentiable. This
set reduces to: (a) the K.P.P. equation for $n = 0$, (b) the
FitzHugh-Nagumo equations with $n = 1$, and (c) the Hodgkin-
Huxley equations with $n = 3$. Writing $W \equiv \mathrm{col}\,(V, w_1, \cdots, w_n)$
and assuming a traveling-wave solution of the form $W(x, t) =$
$W_T(x-ut) = W_T(\xi)$, a general solution can be written $W(\xi, \tau) =$
$W_T(\xi) + W_P(\xi, \tau)$. The linearized equation for $W_P$ is then [as in
(5.1.7)]

$$\begin{bmatrix} V_{P,\xi\xi} \\ 0 \\ 0 \\ \vdots \\ 0 \end{bmatrix} + uW_{P,\xi} - W_{P,\tau} = AW_P \qquad (5.1.43)$$

where $A$ is an $(n+1) \times (n+1)$ matrix with elements obtained by
differentiating the $F$ values with their arguments and evaluating
at $W_T$. Evans shows:

1.  The solution for (5.1.42) decays exponentially to $W_T(\xi + \alpha)$ (from a suitably small initial perturbation) if and only if the solution for (5.1.43) decays exponentially to $W_{T,\xi}$.

2.  The solution for (5.1.43) decays exponentially to $W_{T,\xi}$ if and only if the associated eigenvalue equation

$$\begin{bmatrix} \phi_{0,\xi\xi} \\ 0 \\ 0 \\ \cdot \\ \cdot \\ \cdot \\ 0 \end{bmatrix} + u\Phi_\xi + (s - A)\Phi = 0 \qquad (5.1.44)$$

where

$$\Phi \equiv \mathrm{col}(\phi_0, \phi_1, \cdots, \phi_n)$$

has no eigenvalues with negative real parts; and $\Phi = W_{T,\xi}$ is the <u>only</u> eigenfunction for $s = 0$.

A similar result has quite recently been obtained by Sattinger (1976) for a more general system that allows the F functions in (5.1.42) to depend on the $w_{i,x}$. The zero eigenvalue of the linear operator must be isolated at the origin of the complex plane, and the remaining eigenvalues must lie within a certain parabola in the right half plane. Evans (1975) has extended his work to show that there must be an unstable pulse as well as a stable pulse.

Thus far we have restricted our attention to the notion of <u>temporal instability</u>, that is, an unbounded growth with time of a perturbation that is bounded in space. We might also consider <u>spatial instability</u>: an unbounded growth with space of a perturbation that is bounded in time. For a propagating pulse that is constrained to approach zero as $x \to \pm\infty$, these two notions are the same; the disturbance is bounded in both space and time; and growth of the peak (say) with time is equivalent to growth with space. Rinzel (1975a, b) has recently emphasized that the matter is not so simple for periodic traveling-waves. From a numerical study of perturbations of the periodic waves discussed in connection with Fig. 4-14, he has shown that the conditions for spatial and temporal stability do not necessarily coincide. Furthermore, for "signaling" experiments in which a fiber is stimulated by a periodic disturbance of fixed amplitude at a fixed point (by an electro-physiologist or a sensory input), spatial instability is of primary interest.

The condition for spatial instability is simply stated for the periodic waves in Fig. 4-14 with reference to the corresponding "frequency-wave number" plot shown in Fig. 5-1. Here frequency $f = 1/T$ and wave number is $1/\lambda$, where $T$ and $\lambda$ are, respectively, the temporal and spatial periods of the wave. Evidently the traveling wave velocity is given by

$$u = f\lambda \qquad (5.1.45)$$

The transition from spatial stability to spatial instability occurs at the maximum frequency, $f_{max}$. This is the frequency at which

$$\frac{df}{d(\frac{1}{\lambda})} = 0 \qquad (5.1.46)$$

Larger velocities are spatially stable and smaller velocities, spatially unstable. Rinzel (1975a, b) found temporal instability at $f_{max}$ for $\epsilon = .05$ with several values of $V_1$, as defined in

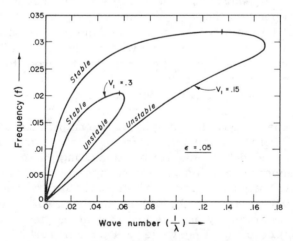

FIGURE 5-1.    Plot of frequency against wave number for periodic traveling-wave solutions of the FitzHugh-Nagumo equation with $F(V)$ defined as in (4.3.21), $\epsilon = 0.05$, $a = 0$, and $b = 0$ (redrawn from Rinzel, 1975).

(4. 3. 21). Furthermore $f_{max}$ does not occur at the minimum value of $\lambda$ so a certain portion of the upper branches of the curves in Fig. 4-14 will be spatially unstable.

The stability investigation of waveforms on myelinated fibers is yet to begin. Beyond the speculations associated with Nasanov diagrams (see Fig. 4-21), there is only the work of Predonzani and Roveri (1968), which treats equilibrium stability of a lossless transmission line that is periodically loaded with active bipoles. Thus much remains to be done before the study of waveform stability is complete.

## 2.   THRESHOLD FOR A SPACE-CLAMPED MEMBRANE

Although we are primarily concerned with the excitation of a propagating action potential on an active nerve fiber, it is interesting to begin with a brief consideration of threshold effects on the space-clamped membrane indicated in Fig. 3-7. A classical threshold experiment is to measure the relation between the strength and the duration of a stimulation just sufficient to induce an action potential on the membrane. For convenience let us suppose that the total membrane area in Fig. 3-7 is 1 cm$^2$ and assume that $I_{12}(t)$ is a square pulse with the form

$$I_{12}(t) = 0 \quad \text{for} \quad 0 > t > \tau$$

$$= I \quad \text{for} \quad 0 < t < \tau$$

The point of the strength-duration measurement is to increase I and/or $\tau$ until the action potential is observed and then record the relation $I(\tau)$.

A rough idea of what should be expected can be found by considering the equivalent circuit for the membrane in Fig. 4-5. This linearized representation should be approximately correct below threshold, where the effects of nonlinearities have not yet become dominant. Since $I_{12}(t)$ is a step function at $t = 0$, it is convenient to use the Laplace transform technique (Gardner and Barnes, 1942) to find the resulting membrane voltage as a function of time. The membrane impedance is

$$z = \frac{I}{C}\left[\frac{s + R/L}{s^2 + Rs/L + 1/LC}\right] \qquad (5.2.1)$$

where  s  is complex frequency.   The Laplace transform of a current step with amplitude  I  is  I/s  so below threshold the Laplace transform of the voltage will be

$$\mathcal{L}[v(t)] = \frac{I}{C}\left[\frac{s + R/L}{s(s^2 + Rs/L + 1/LC)}\right] \tag{5.2.2}$$

and the voltage across the membrane as a function of time can be cbtained by inverting this transform.   If it is assumed that  $V_1$  is a threshold voltage above which the membrane will exhibit an action potential, the strength-duration curve for threshold can be determined by setting

$$\mathcal{L}v(\tau) = V_1 \tag{5.2.3}$$

The limiting cases for long and short values of  $\tau$  are easily determined since large values of  t  correspond to small values of  s  and vice versa.   Thus for short times (5.2.2) becomes approximately

$$\mathcal{L}[v(t)] \approx \frac{I}{Cs^2} \tag{5.2.4}$$

or

$$v(t) \approx \frac{I}{C}t \tag{5.2.5}$$

Setting  v  equal to the threshold voltage  $(V_1)$  and  t  equal to the duration of stimulation  $(\tau)$  as indicated in (5.2.3) yields the relation

$$I\tau \approx V_1 C \tag{5.2.6}$$

for  $\tau \ll L/R$.   The fixed quantity of charge appearing on the right-hand-side of (5.2.6) is the charge that must be supplied to the membrane capacitance in order to change its voltage by an amount equal to the threshold voltage.   If, on the other hand,  $\tau \gg L/R$, then (5.2.2) becomes approximately

$$\mathcal{L}[v(t)] \approx \frac{IR}{s} \tag{5.2.7}$$

or

$$v = IR \qquad\qquad (5.2.8)$$

Requiring that $v = V_1$ in (5.2.8) implies a "rheobase" current equal to $(V_1/R)$. But overshoot in the voltage response to a current step [which was clearly indicated by Hodgkin and Huxley (1952d) and reemphasized by Muro, Conti, Dodge, et al. (1970)] will permit the voltage to reach threshold at stimulating currents less than $(V_1/R)$. Thus $(V_1/R)$ should be considered as the rheobase when current is <u>slowly</u> increased.

This discussion greatly oversimplifies the dynamics involved during excitation of a real active membrane. For a survey of the history of strength-duration measurements and a thorough discussion of current research problems, the reader is referred to Chapters 6 and 7 of Khodorov's <u>The Problem of Excitability</u>, which has recently been published in English. Particular problems which require more careful consideration include the following:

a.  <u>Definition of Threshold Voltage</u>.

As Khodorov (1974) has emphasized, there is not an unambiguous definition for the threshold voltage. In general the time derivative of the membrane voltage is

$$\frac{dv}{dt} = \frac{I_{12} - I_i}{C} \qquad\qquad (5.2.9)$$

where $I_{12}$ is controlled by the experimenter (see Fig. 3-7) and $I_i$ is the ion current that should, in principle, be calculated in some precise way such as through the Hodgkin-Huxley equations (3.2.3). For a very short stimulating pulse, $I_{12} = 0$ when the threshold is reached, so the condition can be defined as

$$I_i < 0 \qquad\qquad (5.2.10)$$

or

$$I_{Na} > I_K + I_L \qquad\qquad (5.2.11)$$

For a very long pulse, however, $I_{12} \neq 0$ and the threshold voltage might be defined as the point where upward curvature of voltage begins. This implies

$$\frac{d^2v}{dt^2} > 0 \qquad\qquad (5.2.12)$$

or

$$\frac{dI_{Na}}{dt} > \frac{dI_K}{dt} + \frac{dI_L}{dt} \qquad\qquad (5.2.13)$$

Defined in this way, the threshold voltage can vary by several millivolts as the strength of the stimulating current is changed.

b.  Accomodation.

    It has long been an established experimental fact that a slow turn-on of the stimulating current leads to the observation of a higher rheobase current. Although this effect is qualitatively predicted by analysis of the simple membrane equivalent circuit in Fig. 4-5, some related effects are not. In particular a "minimum slope" for increase of the current stimulation with time is often observed (Khodorov, 1974). Below this minimum slope no action potential is observed at any level of stimulating current. Typical values for the minimum slope are often of the order of 0.1 rheobase units per millisecond. This effect is predicted by the complete Hodgkin-Huxley equations, and it comes about because with a slowly increasing voltage the sodium turn-off (see $h_0$ in Fig. 3-12) can hold the inward sodium current to a sufficiently low value. There are other adaptation effects (FitzHugh, 1969) with time constants of the order of a second that are not represented at all by the Hodgkin-Huxley equations.

3.  THRESHOLD FOR AN ACTIVE FIBER

    An experimental arrangement for measuring the strength duration curve for threshold excitation of a propagating action potential on a nerve fiber is indicated in Fig. 5-2. Computations by Cooley and Dodge (1966) of strength-duration curves for the Hodgkin-Huxley axon are presented in Fig. 5-3, and these agree well with experimental results (Noble and Stein, 1966; Cole, 1968; Khodorov, 1974). It is important to notice that for small values of the duration $(\tau)$ Fig. 5-3 takes the form

$$I\tau = Q_\theta \tag{5.3.1}$$

where $Q_\theta$ is a constant <u>threshold charge</u> just as (5.2.6). In this section we shall be primarily concerned with the computation and physical understanding of this charge.

To begin let us proceed as in the previous section and represent the shunt admittance per unit length of the fiber below threshold as (Scott, 1973a)

$$y = c \; \frac{s^2 + r_1 s/\ell + 1/\ell c}{s + r_1/\ell} \tag{5.3.2}$$

Then the voltage at the input terminals, $v(0, t)$, will be related to the input current by the characteristic impedance, $Z_O$, of the fiber. This is the square root of the series impedance per unit length divided by the shunt admittance per unit length (Scott, 1970). Thus

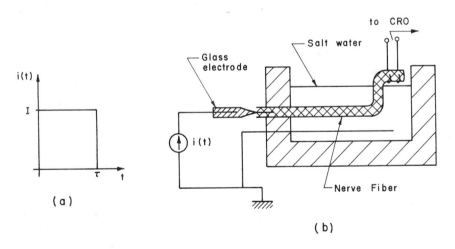

FIGURE 5-2.    (a) Strength (I) and duration $(\tau)$ for a threshold measurement; (b) experiment to measure strength-duration curves for a nerve fiber.

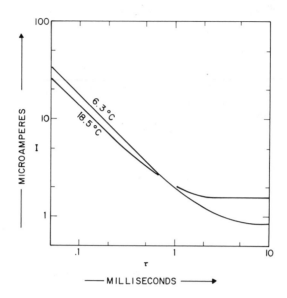

FIGURE 5-3.    Calculated strength-duration curves for the Hodgkin-Huxley axon (Cooley and Dodge, 1966).

$$Z_o \equiv \sqrt{\frac{r_s}{y}} \qquad (5.3.3)$$

$$= \left[ \frac{\dfrac{r_s}{c}\left(s + \dfrac{r_1}{\ell}\right)}{s^2 + \dfrac{r_1}{\ell}s + \dfrac{1}{\ell c}} \right]^{\frac{1}{2}} \qquad (5.3.4)$$

Again the Laplace transform of a current step of amplitude I is I/s, so the transform of the voltage at the input terminals

$$\mathcal{L}[v(0, t)] = I\frac{Z_o}{s} \qquad (5.3.5)$$

The inverse of this transform can be obtained from tables (Roberts and Kaufman, 1966) by convolving

$$\mathcal{L}^{-1}\left[\frac{\sqrt{s+\dfrac{r_1}{\ell}}}{s}\right] \quad \text{with} \quad \mathcal{L}^{-1}\left[\frac{1}{\sqrt{s^2+\dfrac{r_1}{\ell}s+\dfrac{1}{\ell c}}}\right]$$

to obtain

$$v(0,t) = I\sqrt{\frac{r}{c}}\left[\frac{\exp\left(-\dfrac{r_1}{\ell}t\right)}{\pi t} \times \sqrt{\frac{r_1}{\ell}}\ \text{erf}\left(\sqrt{\frac{r_1 t}{\ell}}\right)\right]$$

$$\otimes\left[\exp\left(-\dfrac{r_1 t}{2\ell}\right)J_0\left(\sqrt{\frac{1}{\ell c}-\frac{4r_1^2}{\ell^2}}\ t\right)\right]$$

(5. 3. 6)

where the symbol $\otimes$ denotes convolution or

$$a(t) \otimes b(t) \equiv \int_{-\infty}^{\infty} a(y)b(t-y)dy$$

$$= b(t) \otimes a(t)$$

The error function is defined as

$$\text{erf}(z) \equiv \frac{2}{\pi}\int_0^z e^{-y^2}dy$$

and $J_0$ is the zeroth order Bessel function as defined by Watson (1962).

Now if we attempt to derive a strength duration curve, as in (5. 2. 3), by setting

$$v(0,\tau) = V_1 \qquad (5. 3. 7)$$

the axial current

$$i(0, \tau) = -\frac{1}{r_s} v_x(0, \tau) \tag{5.3.8}$$

is being neglected.  For small values of $\tau$, this means that the sodium component of membrane current must supply not only the potassium and leakage components, as indicated in (5.2.11), but also the axial current.  If $\tau$ is long enough, then $v_x$ will be small at the input end.  This critical value of $\tau$ should be of the order of $\sqrt{\ell c}$ (see Fig. 4-5), which is the time necessary for ion current to begin to flow through the membrane.  Thus (5.3.7) is applicable only for stimulations of duration

$$\tau > \sqrt{\ell c} \tag{5.3.9}$$

To plot solutions of (5.3.7) it is convenient to normalize as

$$\tau_n = \frac{\tau}{\sqrt{\ell c}}$$

$$\tag{5.3.10a, b}$$

$$I_n = I \frac{\sqrt{r_s r_1}}{V_1}$$

and thus the strength–duration curve (5.3.7) becomes

$$I_n = \left\{ \int_0^{\tau_n} \left[ \frac{e^{-Ay}}{\sqrt{\pi A y}} + \mathrm{erf}(\sqrt{Ay}) \right] \left[ \exp(-\tfrac{1}{2}A(\tau_n - y)) J_0(\sqrt{1 - 2A^2}(\tau_n - y)) \right] dy \right\}^{-1} \tag{5.3.11}$$

where

$$A \equiv r_1 \sqrt{\frac{c}{\ell}} \tag{5.3.12}$$

(which is the reciprocal of the membrane "Q") and the restriction (5.3.9) becomes

$$\tau_n > 1 \tag{5.3.13}$$

Strength-duration curves for various values of the constant A in (5. 3. 11) are plotted in Fig. 5-4.   The following points should be noted:

1.  Comparison with Fig. 5-3 shows that (5. 3. 11) is clearly incorrect when the inequality (5. 3. 13) is violated.   Indeed, (5. 3. 11) implies $N\sqrt{\tau}$ = constant for small $\tau$ rather than the condition (5. 3. 1).

2.  The dashed lines in Fig. 5-4 indicate the actual values plotted from (5. 3. 11).   According to (5. 3.7), however, threshold is reached when the highest value of voltage reaches the threshold, $V_1$.   Thus the solid lines in Fig. 5-4 are drawn with the assumption that firing takes place on the voltage "overshoot. "

3.  The observation by Mauro, Freeman, Cooley, et al. (1972) on squid fibers indicates a larger voltage overshoot for a lower temperature.   In terms of the simple membrane equivalent circuit (Fig. 4-5 ), this implies a larger membrane "Q" at lower temperature or, from (5. 3. 12), a smaller value of A.   Thus Fig. 5-4 qualitatively explains the "cross-over" of the threshold curves displayed in Fig. 5-3.

4.  The increase in rheobase for a slowly increasing stimulation is indicated on Fig. 5-4 for the curve A = 0. 3.   If the fiber is stimulated with a step of current, there will be a considerable overshoot and the rheobase will be Rh 1.   If, on the other hand, the stimulating current is slowly raised (e. g. , as a linear ramp function), the overshoot will not occur and the larger rheobase (Rh 2) will be observed.   Of course the caveats outlined in the previous section concerning the qualitative nature of this representation still apply.   In particular, (5. 3. 11) does not predict a minimum gradient for excitation.

But how are we to calculate the strength-duration curve when $\tau$ is short, so that longitudinal current away from the input cannot be neglected and (5. 3. 7) does not apply?   What is the physical significance of the threshold charge $Q_\theta$ that appears in (5. 3. 1)? In order to answer these questions it is helpful to turn our attention briefly toward the threshold problem for the superconductive

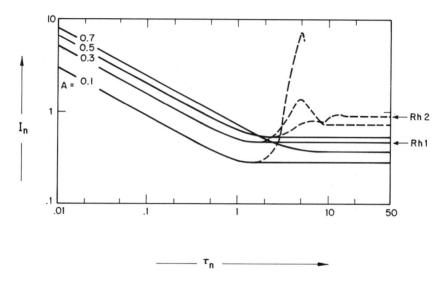

FIGURE 5-4.    Normalized strength-duration curves calculated from
(5. 3. 11), which assumes threshold is achieved when
the input terminal voltage changes by a fixed value
($V_1$).    This assumption is only valid for $\tau_n > 1$.

neuristor which was introduced in Section 4-8.    Equations (4. 8.2a)
and (4. 8. c) can be combined into the <u>conservation</u> <u>law</u>

$$\frac{\partial v}{\partial x} + \frac{\partial (\ell i_1)}{\partial t} = 0 \tag{5. 3. 14}$$

where  $v$  is the <u>flow</u> of the conserved quantity and $\ell i_1$ is its
<u>density</u>.    Thus the conserved quantity is the magnetic flux

$$\Phi = \ell \int_{-\infty}^{\infty} i_1 \, dx \tag{5. 3. 15}$$

From a physical point of view it is not at all surprising to find
magnetic flux conserved in a region between two superconducting
boundaries, but analytically it is rather convenient.    If the stimulation is a voltage pulse of strength  $V$  and duration  $\tau$ , the

magnetic flux introduced by the source is just $V\tau$. To achieve
threshold this flux must be sufficient for the unstable pulse indicated on the horizontal line of Fig. 4-33. Calling the flux of the
unstable pulse $\Phi_P(I_B)$, the threshold condition on strength (V)
and duration $(\tau)$ is simply (Scott, 1973a)

$$V\tau = \Phi_P(I_B) \tag{5.3.16}$$

If this much flux is supplied, the pulse can grow into the stable
leading and trailing edges as indicated in Fig. 4-32. Threshold
levels of strength are plotted against duration in Fig. 5-5 from
measurements on the eighty-two-section analog of (4.8.2) discussed in Section 4-8 (Reible and Scott, 1975). Evidently (5.3.16)
is well satisfied. The decrease in $\Phi_P$ with increasing bias current $(I_B)$ arises because the ratio $A_2 : A_1$ increases (see Figs.
4-7 and 4-31), which reduces the amplitude of the unstable pulse.

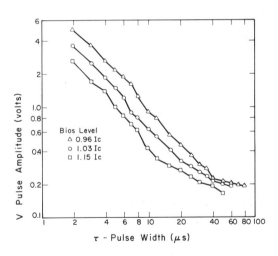

FIGURE 5-5.    Plot of strength (V) against duration $(\tau)$ for threshold excitation of the superconductive neuristor dis-
cussed in Section 4-8. Measurements were made on
the eighty-two-section electronic analog of (4.8.2).

Now we can use the same concept to understand (5. 3. 1). To see this, note that the second Hodgkin-Huxley equation (4. 1. 1b) can be written as the approximate conservation law

$$\frac{\partial i}{\partial x} + \frac{\partial (cv)}{\partial t} \approx 0 \qquad (5.\,3.\,17)$$

as long as the ion current density through the membrane is small compared with the displacement current density. From (4. 1. 1b) this requires

$$|j_i| << |c\,\frac{\partial v}{\partial t}| \qquad (5.\,3.\,18)$$

Inspection of the many computations of threshold stimulation by Khodorov (1974), and in particular the 2. 05-cm point in Fig. 4-25, shows that the inequality (5. 3. 18) is satisfied during the establishment of a threshold pulse on the fiber. For the approximate conservation law (5. 3. 17), i is the _flow_ of the conserved quantity and cv is its _density_. Thus the conserved quantity is electric charge, which is just what we wish to determine for (5. 3. 1). Integrating the flow over time on the leading edge of a threshold pulse gives

$$Q_\theta = \int_{LE} i\,dt \qquad (5.\,3.\,19)$$

which from (4. 1. 1a) can be written

$$Q_\theta = -\frac{1}{r_s} \int_{LE} \frac{\partial v}{\partial x}\,dt$$

Now the threshold pulse is a traveling wave of the form $v(x - u_B t)$, where $u_B$ is the velocity on the lower (unstable) branch of Fig. 4-4. Thus

$$\frac{\partial v}{\partial x} = -\frac{1}{u_B}\frac{\partial v}{\partial t}$$

and

$$Q_\theta = \frac{V_B}{r_s\,u_B} \qquad (5.\,3.\,20)$$

where $V_B$ is the amplitude of the threshold pulse.  The accuracy of this equation is easily checked.  For the Hodgkin-Huxley axon

$$r_s = 1.94 \times 10^6 \text{ ohm/m}$$

whereas from Fig. 4-3 (curve b) at 18.5°C

$$u_B = 5.66 \text{ m/sec}$$

$$V_B = 18 \times 10^{-3} \text{ V}$$

Thus (5.3.20) gives

$$Q_\theta = 1.64 \times 10^{-9} \text{ C at 18.5°C}$$

whereas the corresponding value from the Cooley and Dodge (1966) calculations in Fig. 5-1 is

$$Q_\theta = 1.33 \times 10^{-9} \text{ C at 18.5°C}$$

Considering the simplicity of (5.3.20) this is rather good agreement.

It is interesting to compare the charge required to bring a fiber to threshold with the charge stored in the leading edge of a fully developed action potential, $Q_0$.  Just as in (5.3.20) this leading edge charge can be computed as

$$Q_o = \frac{V_A}{r_s u_A} \tag{5.3.21}$$

where $V_A$ and $u_A$ are the amplitude and velocity of a fully developed action potential.  From Hodgkin and Huxley (1952d) at 18.5°C

$$V_A = 90.5 \text{ mV so that } Q_o = 2.48 \times 10^{-9} \text{ C}$$

$$u_A = 18.8 \text{ m/sec}$$

whereas at 6.3°C

$$V_A = 102.1 \text{ mV so that } Q_o = 4.14 \times 10^{-9} \text{ C}$$

$$u_A = 12.7 \text{ m/sec}$$

From the Cooley and Dodge (1966) calculations in Fig. 5-1

$$Q_\theta = 1.73 \times 10^{-9} \text{ C at } 6.3^\circ\text{C}$$

Thus, in general, we can write

$$Q_\theta = \alpha \, Q_o \qquad\qquad (5.3.22)$$

where for the Hodgkin-Huxley axon

$$\alpha = 0.54 \quad \text{at} \quad 18.5^\circ\text{C}$$

$$= 0.42 \quad \text{at} \quad 6.3^\circ\text{C}$$

This axon carries on the leading edge of a fully developed action potential about twice the amount of charge required to excite a threshold pulse.

　　These considerations are pertinent to the problem of blockage of the action potential at a point of abrupt widening (see Section 4-6). The pulse should fail to pass when the leading edge charge carried into the discontinuity by the smaller fiber is insufficient to supply the threshold charge required by the larger fiber. Since pulse velocities are proportional to $a^{\frac{1}{2}}$ (where $a$ is the fiber radius) and the series resistance is proportional to $a^{-2}$, the leading edge charge in (5.3.21) can be expressed as

$$Q_o = ka^{3/2} \qquad\qquad (5.3.23)$$

where $k$ is a factor that is independent of fiber diameter. Likewise the threshold charge can be written

$$Q_\theta = \alpha k a^{3/2} \qquad\qquad (5.3.24)$$

where the factor $\alpha$ is approximately equal to $\frac{1}{2}$ and relatively insensitive to temperature for the Hodgkin-Huxley axon. However, the leading edge charge that an action potential carries into a discontinuity will be greater than that given in (5.3.23) because, as we saw in Section 4-6, the pulse slows down on approaching an enlargement. Thus the condition for passage of a pulse can be expressed as

$$\left(\frac{a_2}{a_1}\right)^{3/2} < \frac{\zeta}{\alpha} \tag{5.3.25}$$

where $\zeta$ is the factor by which the incoming pulse increases its leading edge charge as it approaches the discontinuity.

It is interesting to compare this result with (4.6.34) obtained by Pastushenko and Markin (1969). To estimate $\zeta$ we can assume that incoming charge is proportional to the pulse amplitude and inversely proportional to its velocity, as indicated in (5.3.21). For a Hodgkin-Huxley axon at $20^{\circ}C$, Berkinblit, Vvedenskaya, Gnedenko, et al. (1970) report:

> The abrupt widening of a nerve fiber (more than 5.5 times) leads to blocking of the impulse. Widening less but close to critical leads to a sharp drop in the amplitude of the a. p. (to 20-30 per cent of the initial value) and sharp slowing of the speed of conduction (from 18 m/s to 2 m/s) which determines the considerable delay in conduction on passage of the impulse through the widening.

which implies $\zeta \doteq 3$. Taking $\alpha = \frac{1}{2}$, (5.3.25) then indicates blocking at a critical ratio greater than 3.3 which is in agreement with the result of Pastushenko and Markin (4.6.34) but somewhat leas than the actual ratio of 5.5.

Assuming $\alpha = \frac{1}{2}$ and a critical expansion ratio of 5.5, (5.3.25) implies $\zeta = 6.44$, which does not appear to be inconsistent with the numerical results of Berkinblit, Vvedenskaya, Gnedenko, et al. (1970). Further studies would certainly be of value. We return to such considerations in Chapter 6 when we consider conditions for blocking of impulses at the branching of an active fiber.

# 6

# Pulse Interactions on the Multiplex Neuron

Visible, invisible,
   a fluctuating charm
an amber-tinctured amethyst
   inhabits it, your arm
approaches and it opens
   and it closes; you had meant
to catch it and it quivers;
   you abandon your intent.

Marianne Moore[*]


   In Chapters 4 and 5 we have considered the nature of propagating nerve impulses, how they interact with nonuniformities of a nerve fiber, and the conditions necessary to induce them. We are now in a position to augment this discussion of level 3 in the scientific hierarchy by investigating the interactions of pulses in and between neurons. Those of us with a background in the physical sciences often underestimate the functional complexity of a single nerve cell, describing it as a simple device that compares a weighted sum of dendritic (input) signals with some "threshold"

[*] From The Complete Poems of Marianne Moore. Copyright © 1959 by Marianne Moore. Reprinted by permission of the Viking Press and of Faber & Faber Ltd.

level above which an output pulse is transmitted along a branching
axon.  No better introduction could be suggested to the variety of
structures exhibited by real nerve cells than a few hours with
Ramón y Cajal's Histologie du Systeme Nerveux.  From this classic work the present author extracts only one drawing, namely,
the Purkinje cell of the human cerebellum shown in Fig. 6-1.  The
vast aborization of the dendritic fibers accepts some 80, 000 synaptic inputs (Eccles, 1973) from "parallel fiber" axons of the
granule cells (a) in Fig. 1-2.  Studies with the electron microscope
(Hamlyn, 1963; Poritsky, 1969) indicate a very complex encrusting
of cell bodies and even axons with synaptic contacts as is indicated in Fig. 6-2.  On the axonal (output) side it is often assumed
that the "parent" fiber excites all "daughters" at each branching
point so that the signal travels without interruption to every distal
(distant) twig, but experiments by Barron and Matthews (1935),
Krnjević and Miledi (1959), Chung, Raymond, and Lettvin (1970),
Parnas (1972), and Grossman, Spira, and Parnas (1973) cast doubt
on this simple picture.  In these studies, the branch points of
some axons emerge as regions of low safety factor where high-
frequency blockage, alternate firing, and other forms of information processing can occur.  Branch-point conductance might be
influenced by small changes in local geometry and electric coupling, thus providing locations for modification of neural transmission or learning.  On the dendritic (input) side of the nerve cell
body, the situation is even less clear.  There are experimental
results indicating that information proceeds through the dendritic
trees of some neurons by purely passive means (Purpura and
Grundfest, 1956; von Euler, Green, and Ricci, 1956; Grundfest,
1958), and a corresponding mathematical theory of passive dendrites has been developed (Rall, 1959, 1962a, b, 1964, 1967;
Pokrovskii, 1970) that essentially involves a linearized diffusion
equation with space dependent coefficients.  But experiments indicating passive dendritic conductance are open to various interpretations (Bishop, 1958; Eccles, 1960; Rall and Shepherd, 1968;
Rall, 1970; Bogdanov and Golovchinskii, 1970), and there have
been several studies implying that action potentials can propagate
at least on the larger branches of some dendritic trees.  Lorente
de Nó (1960), Arshavskii, Berkinblit, Kovalev, et al. (1965),
Llinás, Nicholson, and Precht, (1969), Pastushenko, Markin, and
Chizmadzhev (1969 a, b), Gutman (1971), Berkinblit, Dudzyavichus,
and Chailakyan (1971), Scott (1973b), and Gutman and Shimoliunas
(1973) have pointed out that the dendrites should be able to perform

elementary logical operations at branching points if they can propagate action potentials or even decremental pulses.  In simple terms, the branch may act as a logical "OR" if a pulse on either daughter can supply sufficient charge to excite a pulse on the parent; otherwise, it may act as an "AND".

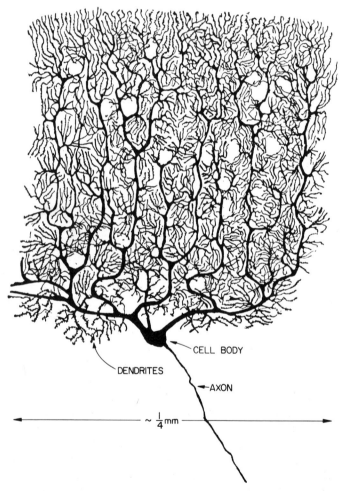

FIGURE 6-1.    Purkinje cell of the human cerebellum (Ramón y Cajal, 1952).

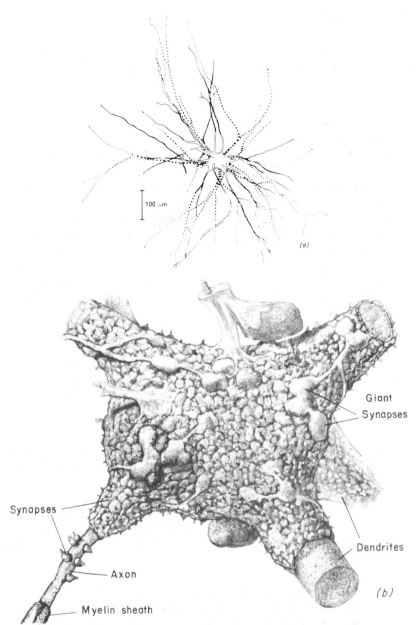

FIGURE 6-2.   (a) Reconstruction of a Procion-dye injected motoneuron by Barrett and Crill (1974a);  (b) synaptic
contacts on the cell body of a cat's motoneuron
(Poritsky, 1969).

Schmitt, Dev and Smith (1976) have recently suggested that the current state of knowledge concerning neuronal circuitry is undergoing a "quiet revolution. " In their words:

> The new view of the neuron, based primarily on recent electron microscope evidence and supported by intracellu-lar electrical recording, holds that the dendrite, far from being only a passive receptor surface, may also be pre-synaptic, transmitting information to other neurons through dendrodendritic synapses. Such neurons may simultaneous-ly be the site of many electrotonic current pathways, in-volving components as small as dendritic membrane patches or individual dendrites. Electrotonic currents, originating in various loci, flow through a vast network; the informa-tion-processing product of these currents is transmitted to other brain regions by projection neurons -- that is, neurons with long axons.

Thus we are led to consider a nerve cell to be at least as complex as the "multiplex neuron" suggested by Waxman (1972) and reproduced in Fig. 6-3. Waxman describes four distinct regions of information processing in a single cell as follows:

1.  The dendritic region in which both excitatory and inhibi-tory synaptic inputs are summed and (possibly) logical decisions are made at branches (shaded).

2.  The nerve body and initial axon segment as shown in Fig. 6-2b. Even the initial segment (or "axon hillock") receives synaptic input to assist in its decision to fire the axon.

3.  The axonal tree, which is often covered by a myelin sheath that restricts membrane current to active nodes and thereby speeds conduction. These nodes can receive inputs and, again, information processing may occur at branches.

4.  The <u>synaptic</u> <u>outputs</u>, which can be modified by input
    contacts from other cells. *

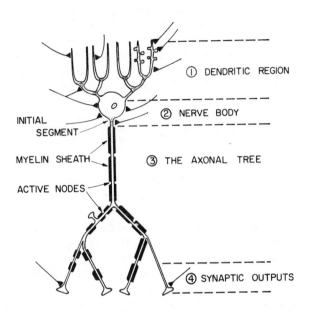

INITIAL
SEGMENT

MYELIN SHEATH

ACTIVE NODES

① DENDRITIC REGION

② NERVE BODY

③ THE AXONAL TREE

④ SYNAPTIC OUTPUTS

FIGURE 6-3.  The multiplex neuron.  Shaded regions have low
             thresholds and may perform logical operations.
             Redrawn with permission of Dr. S. G. Waxman,
             Harvard Medical School; from <u>Brain</u> <u>Res</u>. <u>47</u>:269
             (1972).

In the present chapter we consider some of the ways in which
pulses can interact while propagating along the fibers of a multiplex neuron.  The intent is not to exhaust the subject but to introduce a class of problems that should be of increasing interest
during the next few years.

_____

* In the jargon of integrated circuit technology, a nerve cell may
  be more like a "chip" than a single "gate".

## 1.  SINGLE FIBER INTERACTIONS

The well-established experimental fact that two oppositely directed nerve pulses will annihilate each other on collision is readily understood for the FitzHugh-Nagumo nerve model from our previous development of leading-edge dynamics.  Consider the interaction of two oppositely directed leading-edge transitions shown in Fig. 4-6.  If the approximate conservation law (5. 3. 17) is assumed, then together with (4. 1. 1a) the leading-edge interaction is governed by a linear diffusion equation that can be written

$$\frac{\partial^2}{\partial x^2}(v - V_2) \approx r_s c \frac{\partial}{\partial t}(v - V_2) \tag{6. 1. 1}$$

Thus we expect a relaxation toward $v = V_2$ for $j(v)$ as indicated in Fig. 4-6a if (6. 1. 1) remains valid until the voltage rises above $V_1$.  As soon as $(v - V_2)$ lies within the range of convergence for the Taylor series expansion for $j(v)$ about $V_2$, $v$ must decay to $V_2$.  In terms of (5. 3. 21) we can say that the net approximately conserved charge for the leading edges is zero.  Referring back to Fig. 4-12 for the action potential of the FitzHugh-Nagumo equation, we expect next a slow relaxation with a time constant $\tau_n$ (3. 2.4a). The third stage is the interaction of the trailing edges which, according to the same argument employed for the leading edges, should bring the voltage to a negative value followed by a slow relaxation toward zero.

For the superconductive neuristor which was discussed in Section 4-8, the dynamics of pulse collisions is somewhat more complex.  If, as in Fig. 6-4, $I_B \approx I_C$ (so $A_1 \approx A_2$ in Fig. 4-6), pulse destruction is observed on the electronic analog of (4. 8. 2). If $I_B > 1.2 I_C$, the pulses return to their full amplitude as shown in Fig. 6-5.

## 2.  PARALLEL FIBER INTERACTIONS

No more than a glance at Fig. 4-18b should be necessary to justify an interest in the interactions of pulses on parallel fibers. Indeed, as early as 1882 Hering used nerves from Kaltfrösche (frogs that had been kept in a cellar at about $0^{\circ}C$ for several

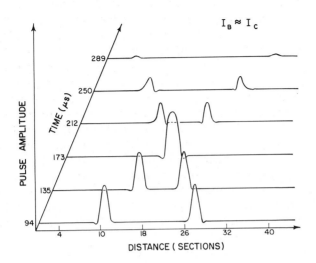

FIGURE 6-4. A destructive collision on the line described in Section 4-8.

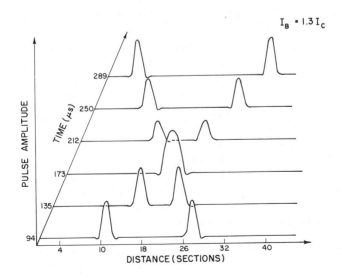

FIGURE 6-5. A nondestructive collision on the line described in Section 4-8.

185

months) to unambiguously demonstrate the excitation of action potentials by those on an adjacent fiber. Since that time, this "ephaptic conduction" has been confirmed by many other investigators (Jasper and Monnier, 1938; Arvanitaki, 1940, 1942; Rosenbleuth, 1941; Renshaw and Therman, 1941) as long as care was taken to enhance the excitability of the second fiber (Granit, Leskell, and Skoglund, 1944). A more subtle effect is the influence on the threshold of a fiber by an action potential on an adjacent fiber (Otani, 1937; Katz and Schmitt, 1939, 1940, 1942; Blair and Erlanger, 1940; Renshaw and Therman, 1941; Marrazzi and Lorente de Nó, 1944; Grundfest and Magnes, 1951). Functionally significant "electrotonic" interaction between giant axons of polychaete worms has been described by Bullock (1953) and between dendrites of electromotor neurons in the mormyrid fish by Bennett, Pappas, Aljure, et al. (1967). Whether such effects are important in the operation of the large cortical mass of cells in the mammalian brain (see Fig. 1-4) remains an open question.

Working with a pair of naturally adjacent, unmyelinated fibers from the limb nerve of a crab (see Fig. 6-6), Katz and Schmitt introduced a reference pulse at AB on fiber ① and at a later time measured the threshold for fiber ② at CD. The result is recorded in Fig. 6-6b and can be interpreted as a stimulation of fiber ② that is roughly proportional to the second derivative of the membrane voltage [or, from (2.30), the membrane current] in fiber ①. To state this point in more physical terms, the total membrane current is outward when the action potential on fiber ① begins to rise; this tends to hyperpolarize fiber ②, which increases its threshold. Near the peak of the action potential, membrane current is inward (mostly $Na^+$) on ①, which tends to depolarize ② and decrease its threshold. Finally on the falling phase of the action potential on ①, its membrane current is outward (largely $K^+$) which tends again to hyperpolarize fiber ②.

Katz and Schmitt also observed the effects of mutual interaction between impulses simultaneously initiated on the two fibers. This effect produced various combinations of speeding or slowing, depending on the phase relation. In particular, synchronization of the pulses could be observed if their independent velocities did not differ by more than about 10%. All interaction effects could be increased by reducing the conductivity of the interstitial fluid. Similar effects have been observed by Crane (1964) on neuristors and by Kunov (1966) on electronic analogs for nerve fibers.

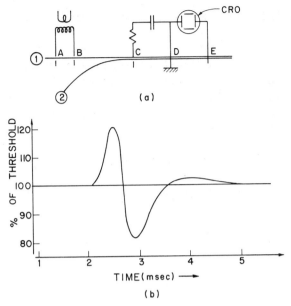

FIGURE 6-6.   (a)  Experiment of Katz and Schmitt (1939) to measure pulse interaction between parallel fibers; (b) change in threshold on ② caused by the presence of a pulse on ① .

Recently Markin (1970a, b) has developed a nonlinear theory for parallel fiber interactions.  Starting from a TLEC representing two unmyelinated fibers that share the external medium (Patlak, 1955), he derived a pair of coupled nonlinear diffusion equations with the form

$$\frac{1}{\gamma}[ (r_2 + r_3)v_{1,xx} - r_3 v_{2,xx}] - c_1 v_{1,t} = j_1$$

$$\frac{1}{\gamma}[ (r_1 + r_3)v_{2,xx} - r_3 v_{1,xx}] - c_2 v_{2,t} = j_2$$

(6.2.1a, b)

where $r_1$, $c_1$, $j_1$, and $v_1$ are the series resistance/length, shunt capacitance/length, membrane ion-current/length, and transmembrane voltage for fiber ① and similarly for fiber ② .  The interstitial resistance/length is $r_3$ and $\gamma \equiv r_1 r_2 + r_1 r_3 + r_2 r_3$; so as

$r_3 \to 0$, (6.2.1) become two uncoupled equations with the form (2.30).

For an analytical study of the interaction effects, Markin uses the Markin-Chizmadzhev (1967) representation for nonlinear pulse propagation which was introduced in Section 4-4. Each fiber is assumed to carry a "piecewise constant" ion current as indicated in (4.4.1) or Fig. 4-15a. The corresponding pulse voltages are resolved into two components as

$$v_1 = v_{11}(\xi_1) + v_{12}(\xi_2)$$
$$v_2 = v_{22}(\xi_2) + v_{21}(\xi_1)$$

(6.2.2a, b)

where

$$\xi_1 = x - u_1 t$$
$$\xi_2 = x - u_2 t + \delta$$

(6.2.3 a, b)

The components $v_{11}$ and $v_{22}$ are the inherent pulses on fibers ① and ② traveling at velocities $u_1$ and $u_2$, respectively. The components $v_{12}$ and $v_{21}$ are the induced voltages from ② to ① and vice versa. The parameter $\delta$ is the distance which the inherent pulse on ② lags behing the inherent pulse on ①.

Assuming $v_{22} = 0$ and making approximations corresponding to those in (4.4.9), Markin (1970a) shows that the stable velocity on ① is

$$u_1 = \sqrt{\left(\frac{J_1}{V_1 c_1^2}\right) \frac{r_2 + r_3}{r_1 r_2 + r_2 r_3 + r_1 r_3}}$$

(6.2.4)

The maximum depolarization potential induced on ② is

$$v_{21}\Big|_{max} \sim \left(\frac{J_2}{J_1}\right)\left(\frac{r_3}{r_2 + r_3}\right)\left(\frac{c_1}{c_2}\right) V_1$$

(6.2.5)

where $J_1$ and $J_2$ are the piecewise constant ion current levels assumed for fiber ① just as in Fig. 4-15a. In a more detailed study that proceeds from electromagnetic theory, Clark and Plonsey (1970) confirm that parameters affecting $r_3$ are of particular importance in determining fiber interaction.   Since both the first two factors appearing on the right of (6. 2. 5) are less than unity, induction of an action potential on ② is not possible for identical fibers $(c_1 = c_2)$.   If the radius of fiber ① is made much larger than that of fiber ② so

$$\frac{c_1}{c_2} \gg 1$$

the situation is not greatly changed.   Markin (1970a) supposes that in the limit of large $a_1$

$$r_3 \approx r_1 \ll r_2$$

so (6. 2. 5) becomes

$$v_{21}\Big|_{max} \sim \left(\frac{J_2}{J_1}\right)\left(\frac{a_2}{a_1}\right) V_1 \qquad (6.\,2.\,6)$$

"Consequently, " he concludes, "the transmission of excitation from one fiber to another is possible only if by virtue of certain factors the threshold of excitation of the second fiber is heavily depressed. "

A key idea in the Markin-Chizmadzhev model for nerve pulse propagation is that conduction velocity is determined by the conditions that raise the leading edge potential to the threshold level $(V_1)$ as indicated in Fig. 4-16.   Markin (1970b) uses this concept to study the synchronization of pulses.   Such synchronization will occur when

$$u_1 = u_2 = u \qquad (6.\,2.\,7)$$

in (6. 2. 3), but this need not be the velocity of a pulse on ① when there is none on ② or vice versa.   The effect of a pulse on ② is to speed up (slow down) a pulse on ① when $v_{12}$ depolarizes (hyperpolarizes) the leading edge.   Either effect can be obtained depending on the distance $\delta$ by which ② lags behind ①.   When $\delta$ is increased from zero, the effect of $v_{21}$ is first to decrease, then increase, then decrease again the velocity of a pulse on ② in a manner which is qualitatively similar to the Katz–Schmitt curve of Fig. 6–6b.  As $\delta$ is decreased from zero, the interaction $v_{12}$ has the same effect on the velocity of pulse ①.  These effects are sketched in Fig. 6–7.  At the intersections of the two curves, (6.2.7) is satisfied and synchronized pulse transmission is possible.

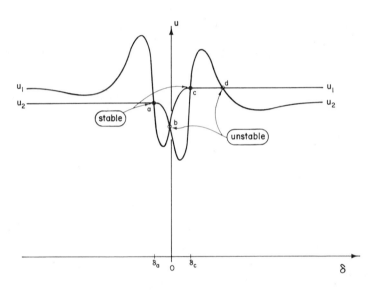

FIGURE 6–7.   Diagram related to the parallel fiber interaction of two Markin–Chizmadzhev pulses (see text for details).

The intersection (c) in Fig. 6–7 occurs at $\delta = \delta_c > 0$, which implies that the pulse on ① is ahead of the pulse on ②.  This is a <u>stable</u> situation because if $\delta$ increases slightly, $u_2$ becomes greater than $u_1$ and $\delta$ tends to decrease.  Intersection (c) occurs at a combined velocity $u_c \approx u_1$ so that the pulse on ① is "pulling" the pulse on ② along at a velocity close to its natural velocity.

The intersection (a) is also <u>stable</u> at $\delta = \delta_a < 0$ which means that the pulse on ② is ahead of the pulse on ①.  The combined velocity $u_a \approx u_2$ so that the pulse on ② is "holding back" the pulse on ①.

In a similar manner, it is readily demonstrated that the intersections (b) and (d) are <u>unstable</u> and hence do not correspond to
experimentally observable pulse synchronizations.

In the nerve fiber bundle of Fig. 4-18b, it is also interesting
to consider the possibility of action potentials traveling in synchronism on <u>several</u> fibers and exciting an additional action potential on an adjacent fiber. This problem has been studied by
Clark and Plonsey (1971) in an electromagnetic analysis that indicates that membrane capacitance plays the primary role in determining the induced transmembrane potential. Less formally,
Markin (1973a, b) supposes that n excited fibers have a combined
capacitance $nc_1$, series resistance $(r_1/n)$, and membrane current
$nj_1$. Then assuming that the series resistance of the unexcited
fiber, $r_2$, satisfies the inequalities

$$r_2 \gg r_3$$

$$\gg r_1/n$$

and also $v_2 \ll v_1$, (6.2.1) assume the form

$$\frac{v_{1,xx}}{r_1 + nr_3} - c_1 v_{1,t} = j_1$$

$$\frac{1}{r_2} v_{2,xx} - c_2 v_{2,t} = j_2 - \frac{\epsilon}{r_2}$$

(6.2.8a, b)

where

$$\epsilon \equiv \frac{nr_3}{r_1 + nr_3}$$

(6.2.9)

All n fibers that propagate synchronized action potentials are represented by (6.2.8a), whereas (6.2.8b) represents a single fiber,
the stimulation of which is being considered. Choosing the fiber
radii to be $10\mu$ and assigning membrane parameters corresponding
to those of the squid (for lack of more appropriate data), Markin
shows that stimulation of fiber ② should occur for $\epsilon \approx 0.5$ or,
from (6.2.9), $n \approx (r_1/r_3)$. If the radius of fiber ② is increased,

$r_2$ becomes smaller and the coupling term in (6.2.8b) increases correspondingly.

The theory of coupled nonlinear diffusion equations can also be applied to study wave propagation in cardiac tissue. Two components of propagation can be taken as activity on both muscle and Purkinje fibers. Each of these activities can be represented as solutions of nonlinear diffusion equations that are strongly coupled through close packing of the two types of fibers. Assuming different refractory periods $(\tau_2)$ for the two waves, Markin and Chizmadzhev (1972) show that stimulation by three pulses (which are separated by a time less than one refractory period and greater than the other) should induce the propagation of a "reverberator." This reverberator has the following properties: (a) it travels at a velocity much less than that of a normal coupled wave, (b) it emits coupled waves periodically in both the forward and backward directions, and (c) it is destructable only through symmetric collision with another reverberator. Markin and Chizmadzhev suggest that the reverberator may be related to fibrillation states of the heart (see also Tsetlin, 1973).

## 3.  CONDUCTION AT BRANCHING POINTS OF AXONS

Let us consider first the situation shown in Fig. 6-8 where an axonal "parent" fiber of radius $a_3$ bifurcates into "daughters" of radii $a_1$ and $a_2$. What will happen to an action potential on the parent when it reaches the branch point? In answering this question it is useful to return to Rall's analysis of a tapered fiber discussed in Section 4-6. He showed (Rall, 1962a) that if the spatial variable is transformed as

$$z = z(x) \tag{6.3.1}$$

where $z(x)$ is determined by

$$\frac{dz}{dx} = a^{-\frac{1}{2}} \left[ 1 + \left( \frac{da}{dx} \right)^2 \right]^{\frac{1}{4}} \tag{6.3.2}$$

the PDE for pulse transmission is invariant as long as the dependence of the fiber radius on $x$ satisfies

$$a^{3/2} \left[ 1 + \left( \frac{da}{dx} \right)^2 \right]^{\frac{1}{4}} = const \qquad (6.3.3)$$

and, more generally, for n fibers that

$$na^{3/2} \left[ 1 + \left( \frac{da}{dx} \right)^2 \right]^{\frac{1}{4}} = const \qquad (6.3.4)$$

For a branching fiber (as in Fig. 6-8) $(da/dx)$ is zero everywhere except at the branch point where it is undefined. Neglecting fields associated with this discontinuity, condition (6.3.4) becomes

$$na^{3/2} = const \qquad (6.3.5)$$

for an undisturbed continuation of the PDE across the branch. This is reasonable from a physical point of view because, as was pointed out in connection with (5.3.3), the <u>characteristic admittance</u> $Y_O$ of a fiber is (Scott, 1970)

$$Y_O = \left[ \frac{shunt\ admittance/length}{series\ impedance/length} \right]^{\frac{1}{2}} \qquad (6.3.6)$$

Shunt admittance/length is proportional to fiber radius; and, if internal resistance dominates, series resistance/length is inversely proportional to radius squared. Thus (6.3.6) can be written

$$Y_O = ma^{3/2} \qquad (6.3.7)$$

where m is a factor which is independent of fiber radius. The condition (6.3.5) requires that characteristic admittance be undisturbed across the branch.

In a recent numerical study, Goldstein and Rall (1974) emphasize the importance of geometric ratio (GR) at a branch. For Fig. 6-8 it is defined as

$$GR = \frac{a_1^{3/2} + a_2^{3/2}}{a_3^{3/2}} \qquad (6.3.8)$$

and, more generally, as the sum of the outgoing characteristic admittances divided by the sum of the incoming characteristic admittances.  For GR = 1, an action potential proceeds through the branch in an undisturbed manner.  For GR < 1, longitudinal current is constricted as the pulse approaches and it accelerates on approach as indicated in Fig. 4-24b.  For GR > 1, the pulse slows down on approach as in Fig.  4-24a and blocking of conduction is possible.

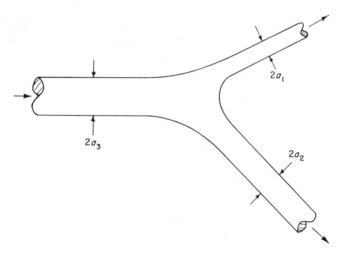

FIGURE 6-8.   Geometry of a branching axon.

    To investigate the conditions for block, we can use the considerations of Section 5-3.  If the leading edge charge carried into the branch by the action potential on the parent is insufficient for the thresholds of the daughters, block will occur.  The incoming charge is

$$Q_i = \zeta k a_3^{3/2} \tag{6.3.9}$$

where $\zeta$ is the factor by which leading edge charge increases as it approaches the branch.  If $u_A'$ and $V_A'$ are, respectively, the changed values of pulse velocity and amplitude as the pulse approaches the branch, then from (5.3.21)

$$\zeta = \left( \frac{V'_A}{V_A} \cdot \frac{u_A}{u'_A} \right)_{max} \tag{6.3.10}$$

Blocking of a Hodgkin-Huxley axon occurs at a critical expansion ratio of 5.5, which, from (5.3.25), implies

$$\zeta \approx 6.44$$

as we noted in Section 5-3.

The condition for block of a pulse into daughter ① in Fig. 6-8 can be calculated by noting that the fraction of input charge $Q_i$ which enters daughter ① is equal to

$$\frac{Y_{01}}{Y_{01} + Y_{02}}$$

the ratio of the characteristic admittance of daughter ① to the total characteristic admittance of all the daughters. From (5.3.24), this charge must be less than

$$\left( \zeta k a_3^{3/2} \right) \left( \frac{m a_1^{3/2}}{m a_1^{3/2} + m a_3^{3/2}} \right) < \alpha k a_1^{3/2}$$

for block. This condition can be written

$$\frac{\zeta}{\alpha} < \left( \frac{a_1}{a_3} \right)^{3/2} + \left( \frac{a_2}{a_3} \right)^{3/2} \tag{6.3.11}$$

and it is the same as the condition for block into daughter ②.

For daughters of same radius this blocking condition becomes

$$\frac{a_3}{a_1} < \left( \frac{2\alpha}{\zeta} \right)^{2/3} \tag{6.3.12}$$

Taking $\alpha = \frac{1}{2}$ and $\zeta = 6.44$ gives the condition $a_3 < 0.28\,a_1$, which is not usually satisfied for a real axon. Thus it can be concluded that isolated pulses should propagate outward without block on an axonal tree.

Under normal physiological conditions, however, the axons usually transmit pulse trains rather than individual pulses. As the calculations by Rinzel and Keller (1973) in Fig. 4-14 indicate, pulse trains of sufficiently high frequency will block even on a uniform fiber. At frequencies approaching this maximum, a GR even slightly greater than unity should lead to blocking of certain pulses at a branch. Such effects have been observed by Barron and Matthews (1935), Krnjević and Miledi (1959), Chung, Raymond, and Lettvin (1970), Parnas (1972), Grossman, Spira, and Parnas (1973). Although this simple theoretical picture implies that all daughters will fire or fail together, it is based on several idealized assumptions (both daughters of circular cross section with negligible external component of series resistance, etc. ), which are probably not valid for branches in a real axonal tree. Once a particular daughter has fired, of course, it will be less sensitive to an immediately following stimulation. Thus there are several possible explanations for preferential firing among the daughters.

Chung, Raymond, and Lettvin (1970) suggest a functional significance for partial conduction through an axonal tree. If only a subset of the distal branches are activated by a single pulse, the axonal tree could translate complex temporal messages into spatial patterns. They conclude as follows.

Several important shifts in perspective stem from the recognition of the complexity of the process of axonal conduction in arborizations and the possible significance such conduction has in spatially structuring interspike interval patterns. Among the most obvious is that "spontaneous" activity and bursty discharge ought not to be regarded as "noise". It is not obvious what any neuron is trying to say, and given the possibility that burstiness may itself be meaningful, we have no basis a priori to decide what is noise and what is message. To do so would imply a prior knowledge of the intentions of the system and its modes of operation or, to use von Neumann's phrase [1958], "the language of the brain".

That areas of low safety factor are very sensitive to extracellular currents raises a second issue. The points of low safety factor present in branched axons imply that neighboring regions in the central nervous system must interact profoundly. The notion that cross talk will degrade the performance of the nervous system is not necessarily true. The degree of interaction that exists suggests that information handling may be aided rather than hampered by cross communication.

Finally, if every pulse arriving at a cell embodies in its spatial distribution an instantaneous statement about the recent history of events, the physiological basis exists for a kind of "short-term memory". Moreover, there is the intriguing possibility that the relative diameters of branches might be structually altered by activity, thereby rendering the system capable of imbedding prior experiences.

## 4.  SYNAPTIC TRANSMISSION

On the schematic diagram of a multiplex neuron in Fig. 6-3, the axonal endings are indicated as enlarged "synaptic outputs" which make contact with other neurons. Some appreciation for the variety of such connections in real tissue may be had from a glance at the Mauthner cell of the goldfish (Fig. 4-22) or the motoneuron of the cat (Fig. 6-2b). Since the advent of the electron microscope, these structures are becoming increasingly well under-stood and several excellent references are available. Katz's Nerve, Muscle, and Synapse is recommended as a clear and pro-vacative introduction to The Physiology of Synapses by Eccles. In addition to presenting a comprehensive review of research re-sults up to 1963, Eccles's book includes an interesting historical survey of the competition between chemical and electrical theories for synaptic transmission. More recent research is discussed in the book by McLennan (1970).

A schematic diagram for a synaptic connection is sketched in Fig. 6-9. Purely electrical transmission of an action potential from the axon ending to the region extending outward from the postsynaptic membrane should be difficult in view of the extreme widening and the fact that a double layer of membrane must be de-polarized in the region of the synaptic cleft. Chemical effects contribute in the following way. The end bulb of the axon stores a large quantity of synaptic vesicles ($\sim 500$ Å in diameter) containing chemical substances, such as acetylcholine and nor-adrenaline, with the ability to selectively alter ionic permeability in the postsynaptic membrane. With reference to Fig. 3-8, an increase in sodium ion permeability $(G_{Na})$ tends to depolarize the postsynaptic membrane and induce transmission of an action potential, whereas an increase in potassium ion permeability $(G_K)$ hyperpolarizes the postsynaptic membrane and inhibits transmission through the synaptic junction. One "quantum" of depolarization or hyperpolarization is associated with the release of chemical transmitter substance from a single synaptic vesicle.

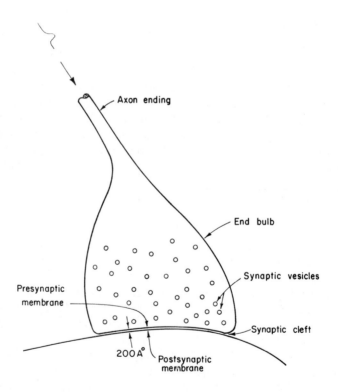

FIGURE 6-9.   Schematic diagram of a synaptic contact.

If the synapse depolarizes the postsynaptic membrane, it is said to induce an "excitatory postsynaptic potential" (EPSP). In the motoneuron, the EPSP appears with a time delay of about 0. 3 msec that is attributed to the liberation of a chemical transmitter substance and its diffusion across the synaptic cleft. After this delay, the EPSP rises to a maximum in about 1-1. 5 msec and then decays exponentially with a time constant of about 5 msec.

Hyperpolarization of the postsynaptic membrane is described as an "inhibitory postsynaptic potential" (IPSP). In a motoneuron, the typical initial delay is about 1. 5 msec, the initial rise is 1. 5 - 2 msec and the time constant for exponential decay is about 3. 3 msec.

But several variations on this simple description of synaptic transmission have been observed. For example, Robertson, Bodenheimer, and Stage (1963) have reported results of an electron microscope examination of the Mauthner cell of the goldfish (see Fig. 4-22) that indicate the possibility of electrical transmission at the "club endings. " Furshpan and Potter (1959a) have studied electrical transmission through the giant motor synapse of the crayfish using glass microelectrodes as indicated in Fig. 6-10a. The time delay observed was very small (usually about 0. 1 msec) which is an indication of electrical transmission. Their observations could be explained by assuming the contact area to function as the diode rectifier indicated in Fig. 6-10c, and they could not be explained by assuming a chemical mechanism. This electrical transmission was distinguished from a slow IPSP (Furshpan and Potter, 1959b) which appears to be chemical in nature.

Transmission across the septal (dividing) membranes (see Fig. 6-10a) of the crayfish lateral giant axon appears to be entirely electrical (Watanabe and Grundfest, 1961). The voltage current characteristic is linear at least over the range $\pm 25\,mV$ and may be represented by a simple series resistor, $R_s$. Transmission of impulses across a septated axon has been investigated by Markin and Pastushenko (1973) using the analytic technique outlined in Section 4-4. They show that the pulse speeds up while approaching the septum and slows down on leaving, as might be expected by comparison with Fig. 4-24b. The net pulse delay is approximately

$$\tau \approx \frac{R_s}{r_s u} \qquad (6.4.1)$$

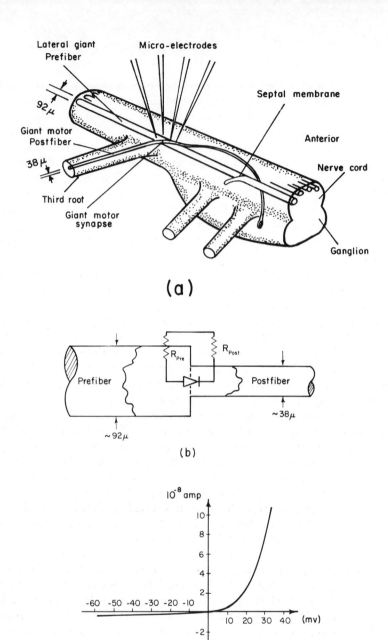

**(a)**

**(b)**

**(c)**

FIGURE 6-10.   (a) Diagram showing the giant motor synapse of the
crayfish; (b) rectifier-equivalent circuit; (c) volt-
ampere characteristic of the rectifier (redrawn from
Furshpan and Potter, 1959b).

for small values of $R_s$.  For a typical crayfish giant lateral axon
with a diameter of $100\mu$, $R_s = 3 \times 10^5 \Omega$, $r_s = 1.2 \times 10^8 \Omega/m$ and
$u = 15$ m/sec so $\tau$ is calculated from (6.4.1) to be about 0.17
msec.  This is compared (Markin and Pastushenko, 1973) to an
observed value of about 0.1 msec.  The difference in these two
values can be partially explained by the neglect of displacement
current through the membrane in deriving (6.4.1).  Markin and
Pastushenko also show that the critical value of $R_s$ for propagation through the septum may be overestimated by a factor of two
from dc arguments.

   Finally, Furakawa and Furshpan (1963) have observed an electrical inhibition of the axon hillock region of the Mauthner cell
(see Fig. 4-22) by the "spiral synapse, " which is followed by a
slower chemical inhibition i. e. , (IPSP).

   Thus it appears that a variety of electrical (fast) and chemical
(slow) output mechanisms are available to the multiplex neuron,
which may be either excitatory or inhibitory in nature.


5.   LINEAR DIFFUSION ON DENDRITES

   The dendritic trees of a multiplex neuron are in a position to
sense many complex spatiotemporal signal patterns on the input
synapses.  Just how a particular neuron recognizes and responds
to a particular input pattern is as yet unclear, but a glance at the
dendrites of the Purkinje cell in Fig. 6-1 should convince the
reader that it may be a rather sophisticated process.  In principle,
each branch should be described by a nonlinear diffusion equation
(2.30) with appropriate continuity conditions at the branching
points and 80,000 synaptic inputs appearing throughout the trees.
Clearly, some simplifying assumptions must be introduced in order
to proceed with an analytical description.  The danger in any simplification, of course, is that the essence of the object under
study may be lost in the quest for a simple model.

   Nonetheless, an interesting possibility is to describe the
branches by linear diffusion equations of the form

$$v_{xx} - r_s c\, v_t = r_s g v \qquad (6.5.1)$$

where $g$ is the conductance per unit length of a dendritic branch
below its firing threshold.  This is essentially the assumption
that action potentials do not develop on dendritic membranes.  As

Rall and Rinzel (1973) have recently emphasized, this may be a reasonable assumption for the cat motoneuron (Fig. 6-2) since the combination of two typical EPSPs yields a response which is equal to or somewhat less than the sum of the individual responses (Burke, 1967; Rall, 1967). Such a sublinear response does not suggest the onset of an action potential, and it can be explained as an interference between chemical depolarizations at adjacent regions of postsynaptic membrane (Kuno and Miyahara, 1969).

Even with the assumption of a linear membrane, however, the analysis of a particular dendritic structure is a rather difficult problem. The basic reason for this additional difficulty is that reflections may occur at each of the branching points. The sum of all signals (reflected, reflected, and reflected again) must be accounted for to obtain an analytic solution. As Rall (1962a, b, 1964, 1967) has suggested, the reflection problems disappears if it is assumed that characteristic admittance is continuous across a branch from daughters to parent and also that all the daughter branches are stimulated in unison. As a simple example, consider the bifurcation sketched in Fig. 6-11. If the sum of the characteristic admittances for branches A and B equals the characteristic admittance of branch C, and if

$$i_A(t) = i_B(t) \tag{6.5.2}$$

then A and B can be considered as components of a single branch that is continuous and uniform across the crotch. If the external component of series resistance can be neglected in comparison with the internal component, then $r_s$ is inversely proportional to radius squared (2.19) and characteristic admittance is proportional to $a^{3/2}$ as was discussed above in connection with (6.3.7). The condition on the radii in Fig. 6-11 is then $2a_1^{3/2} = a_2^{3/2}$. More generally, when the dendritic trunk (of radius $a_t$) has branched into n daughters of radius a,

$$na^{3/2} = a_t^{3/2} \tag{6.5.3}$$

is the condition for reflectionless diffusion.

Recently Rinzel and Rall (Rinzel and Rall, 1974; Rinzel, 1976) have indicated how the reflectionless response of the parent (C) to stimulation of a <u>single</u> daughter (say A) may be obtained. They assume a symmetrical bifurcation as in Fig. 6-11 and suppose first that

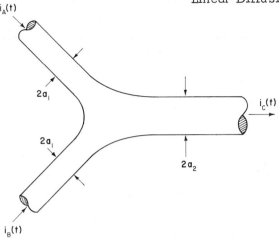

$i_A(t)$

$2a_1$

$2a_1$

$i_B(t)$

$i_c(t)$

$2a_2$

FIGURE 6-11.   Geometry of a symmetrical branching dendrite.

$$i_A(t) = i_B(t) = \frac{1}{2} i_i(t) \qquad (6.5.4)$$

Then, with the admittance matching condition (6.5.3), the corresponding current is readily calculated from the linear diffusion (6.5.1) throughout the branch, and, in particular, $i_C(t)$. Next they assume an antisymmetric stimulation as

$$i_A(t) = - i_B(t) = \frac{1}{2} i_i(t) \qquad (6.5.5)$$

and again calculate current throughout the branch.   Evidently, the antisymmetry condition requires the current to be zero along branch C in this case.   Since the PDE (6.5.1) is linear, the response to an input that is the sum of those in both (6.5.4) and (6.5.5) will be the sum of the individual responses.   But the sum of the symmetric and antisymmetric inputs yields

$$i_A(t) = i_i(t) \quad \text{and} \quad i_B(t) = 0 \qquad (6.5.6a,b)$$

while the sum of the responses on branch C is just the one that was computed for the symmetric input (6.5.4).

Rall and Rinzel (1973) discuss some of the evidence for assuming that the sum of $a^{3/2}$ for daughters equals $a^{3/2}$ of the parent. For the motoneuron the "geometric ratio" of these quantities seems to lie within the range 0.8 to 1.2 with a gradual decrease toward the tips of the tree.   It should be mentioned, of course, that external resistance probably cannot be neglected for the

closely packed cells of real nervous tissue so a modified definition of this ratio may be more appropriate. But assuming, with
Rall and Rinzel, cylindrical fibers and a geometric ratio of unity
at each branch point, (6.5.1) can be normalized as follows. Time
is measured in units of the <u>membrane response time</u>

$$\tau = \frac{c}{g} \qquad (6.5.7\,a)$$

so the dimensionless time variable is

$$T \equiv \frac{gt}{c} \qquad (6.5.7b)$$

Distance is measured in units of the <u>diffusion length</u>

$$\lambda = [r_s g]^{-\frac{1}{2}}$$

Since this changes with position along the dendritic fiber, an appropriate dimensionless space variable is generated by the differential relation

$$dX = \frac{dx}{\lambda}$$

The diffusion length $\lambda \propto a^{\frac{1}{2}}$ and, from (6.5.3)

$$a^{1/2} = n^{-1/3} a_t^{1/2}$$

Thus the dimensionless space variable can more conveniently be
defined as

$$dX \equiv \frac{n^{1/3} dx}{\lambda_t} \qquad (6.5.8)$$

where

$$\lambda_t = [r_{st} g_t]^{-\frac{1}{2}} \qquad (6.5.9)$$

is the diffusion length at the trunk of the dendritic tree. Then the
differential change of membrane area with $x$ is

$$d(\text{Surface area}) = 2\pi a n \, dx = 2\pi a_t \lambda_t \, dX$$

or the circumference of the trunk multiplied by $\lambda_t dX$. With this normalization, (6.5.1) reduces to

$$V_{XX} - V_T = V \qquad (6.5.10)$$

and the "electrotonic length," $\Lambda$, of a dendritic tree can be defined as

$$\Lambda \equiv \int_{\text{base of trunk}}^{\text{end of twigs}} dX$$

For a typical motoneuron the electrotonic length of a dendritic tree appears to lie between 1 and 2 with an average of about 1.5 (Barrett and Crill, 1974a).

Assuming linearity, it is possible to express the voltage response at some point in the dendritic structure (say $v_2(t)$ at point ② ) to the current input at some other point (say $i(t)$ at point ①) as the Green's integral

$$v_2(t) = \int_{-\infty}^{t} i_1(t - t') H_{12}(t') dt' \qquad (6.5.11)$$

where $H_{12}(t)$ is the voltage response at ② to a unit impulse of current injected at ①. The task is to calculate $H_{12}$ between points of interest. Rinzel and Rall (1974) do this in general for the dendritic model indicated in Fig. 6-12, where:

N  is the # of dendritic trees;

M  is the # of symmetrical branchings of each tree;

$\Lambda$  is the "electrotonic" length of each tree.

The electrotonic lengths of the branches are assumed equal. Each branch is assumed to be symmetric and satisfying (6.5.3) so the "reflectionless" calculation discussed above (in connection with Fig. 6-11) can be employed throughout. The membranes of the nerve body and the axon are neglected in comparison with the dendritic membranes.

Of particular interest is the voltage response at the nerve body (point ② in Fig. 6-12) to a unit impulse of current at the

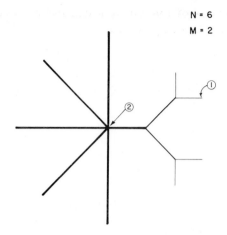

FIGURE 6-12.    General dendritic model for passive-response
calculation.

end of a single branch (point ①).    For these points Rinzel and
Rall obtain

$$H_{12}(t) \doteq \left( \frac{r_{st} \ \lambda \ e^{-t/\tau}}{N\sqrt{\pi} \ t \ \tau} \right) \sum_{n=-\infty}^{\infty} \exp\left\{ -\frac{[(2n-1)\Lambda]^2 \tau}{4t} \right\}$$

$$(6.5.12)$$

In making this calculation (see Fig. 6-12), each of the $N-1$
unstimulated trees is considered as a uniform cylinder of length
$\Lambda$ in the normalized space variable $X$.    The trunk of a dendritic
tree is characterized by:

$r_{st}$ = the series resistance per unit length;

$c_t$ = the capacitance per unit length;

$g_t$ = the membrane conductance per unit length.

The membrane time constant $\tau$ defined in (6.5.8) is $(c_t/g_t)$.
The series (6.5.12) is convenient when $t \to 0$.    For larger
values of $t$ it becomes

$$H_{12}(t) \doteq \frac{e^{-t/\tau}}{N\Lambda\lambda\, c_t} \left\{ 1 + 2 \sum_{n=1}^{\infty} (-1)^n \exp\left[ -\left(\frac{n\pi}{\Lambda}\right)^2 \frac{t}{\tau} \right] \right\} \qquad (6.5.13)$$

The physical implication of (6. 5. 13) can be appreciated by supposing that

$$i_1(t) = Q_0\, \delta(t)$$

where $\delta(t)$ is a unit impulse function (or a Dirac "delta function" of unit area). Then from (6. 5. 11)

$$v_2(t) = Q_0\, H_{12}(t) \qquad (6.5.14)$$

Using the approximate expression (6. 5. 13) for $H_{12}(t)$ we find

$$v_2(t) = \frac{Q_0}{N\Lambda\,\lambda\, c_t}\, e^{-t/\tau}[1 + R(t)] \qquad (6.5.15)$$

where

$$R(t) \equiv 2 \sum_{n=1}^{\infty} (-1)^n \exp\left[-\left(\frac{n\pi}{\Lambda}\right)^2 \frac{t}{\tau}\right] \qquad (6.5.16)$$

Since $N\Lambda\lambda\, c_t$ is just the total membrane capacitance of all the dendrites, (6. 5. 15) implies that the input charge will be evenly distributed over the dendrites as soon as $R(t) \ll 1$. As we noted above, $\Lambda \sim 1.5$ is a reasonable electronic length for the dendritic tree of a motoneuron. Then inspection of (6. 5. 16) (plus the numerical studies of Rinzel and Rall) indicate a uniform charge distribution for $t > \tau$. More generally, we can begin to neglect $R(t)$ in (6. 5. 15) when

$$t > \left(\frac{\Lambda}{\pi}\right)^2 \tau \qquad (6.5.17)$$

For such times, the effect of a charge, $Q_0$, briefly introduced at the twig of a dendritic tree will be to induce a voltage at the cell body equal to

$$v_2(t) = \frac{Q_0}{C_{total}} \cdot e^{-t/\tau} \qquad (6.5.18)$$

where $C_{total}$ is the membrane capacity of all the dendrites.  For small values of time we must return to (6. 5. 12), which implies $H_{12}(t) \to 0$ as $t \to 0$.

It may be helpful at this point to recapitulate the assumptions made in the derivation of (6. 5. 13).  These were:

1.  A geometric ratio of unity is preserved at each branching as implied by (6. 5. 3),

2.  The surface area of the cell body is much smaller than that of the dendrites,

3.  Input current is injected at a twig (distal branch) of a dendritic tree, and, of course,

4.  Membrane voltage evolves according to the linear diffusion equation (6. 5. 1).

For more realistic dendritic geometries, the analytic expressions for response to an impulse of synaptic current can be quite complex.  Many additional analytic and numerical results are available in the studies by Jack and Redman (1971 a, b) on the transient response of a single fiber loaded with a parallel R-C circuit at one end to simulate the nerve body and stimulated at an arbitrary point.  The recent book by Jack, Noble, and Tsien (1975) provides an excellent summary of these calculations.  Butz and Cowan (1974) have developed a simple graphical calculus that generates analytic expressions for the Laplace transform of the impulse response [ $H_{ij}$ defined in (6. 5. 11)] for arbitrary dendritic and cell body geometry.  This calculus should facilitate an automatic computation of linear dendritic response.

The application of such models to real motoneurons is discussed in detail by Jack, Miller, Porter, et al. (1971) and also by Barrett and Crill (1974a).  The latter investigators pay particular attention to the fact that the EPSP induced on a dendrite is not proportional to the time course of the conductance change, $G(t)$, of the postsynaptic membrane (Barrett and Crill, 1974b).  To appreciate this effect, consider that the induced current is given by

$$i_1(t) = G(t)[V_0 + v_1(t)] \qquad (6. 5. 19)$$

where $V_0$ is the difference between the resting potential across the membrane and the diffusion potential [ i. e. , $(V_R - V_{Na})$ for sodium ions as in (4. 1. 12)]. The EPSP and the injected current are $v_1$ and $i_1$, respectively. Assuming an impulse response, $H_{11}(t)$, calculated for the point of current injection, (6. 5. 11) becomes

$$v_1(t) = \int_{-\infty}^{t} G(t')[V_0 + v_1(t')] H_{11}(t - t')dt' \tag{6.5.20}$$

This is a linear Volterra integral equation for $v_1(t)$ (Rinzel and Rall, 1974), which implies that the EPSP will be _less_ than that which would be calculated under the assumption*

$$v_1 << V_0 \tag{6.5.21}$$

To account for this effect Barrett and Crill (1974b) define a "charge injection factor"

$$J \equiv \frac{\int_{-\infty}^{\infty} G(t)[V_0 + v_1(t)] dt}{\int_{-\infty}^{\infty} G(t) V_0 dt} \tag{6.5.22}$$

which is the ratio of charge injected to that which would be under (6. 5. 21). At a distal branch, the effect of a quantal conductance change due to the discharge of a single synaptic vesicle of chemical transmitter into the postsynaptic membrane (see Fig. 6-9 ) is estimated to yield an EPSP of 16-20 mV with a charge deficit $(1 - J)$ of $14 - 19\%$.

If the inequality indicated in (6. 5. 21) is not satisfied, $v_1$ may approach the threshold for an action potential on the dendritic membrane. When this occurs (6. 5. 1) and (6. 5. 20) will no longer be valid because membrane conductance will depend on $v$ as we discussed in Section 3-2, but the qualitative effect will be regenerative $(J > 1)$ with increasing $v$ leading to increasing $G$ leading to increasing $v, \cdots$, as in Fig. 3-11. This situation will be considered in the following section.

---

* Note that $V_0$ is negative while $v_1$ is positive.

## 6.  NONLINEAR INTERACTIONS AT BRANCHING POINTS OF DENDRITES

To find solutions for nonlinear dendritic models that correspond to the linear diffusion calculations discussed in the previous
section would be a difficult task as Pickard (1974) has recently
emphasized.  But it is interesting to consider the suggestion
(Lorente de Nó, 1960; Arshavskii, Berkinblit, Kovalev, et al. 1965)
that the branch points of active dendrites may serve to process the
information carried by dendritic action potentials.  The previous
section and this one correspond, in a way, to the two main sim-
plifications of modern electronics:  linear system analysis and
switching theory.  In the first approach, nonlinearity is dealt with
by assuming that it does not exist, and often this is arranged to be
so.  In the case of switching theory, nonlinearity is desired for
the increased technological possibilities it introduces.  The nonlinearity is rendered analytically tractable by assuming it to be so
strong that dependent variables can take only one of two states.
Then the two element field of Boolean algebra can be used to describe those response functions that are of interest.  Here we discuss the possibility that dendritic branchings can provide the elementary "AND, " "OR, " and "NOT" functions necessary for the
synthesis of an arbitrary Boolean function.  But the present author
does not wish to leave the reader with the impression that Sections
6-5 and 6-6 present antagonistic theories for dendritic function
one of which must eventually be proven "true" and the other "false."
Rather, they should be considered as polar extremes on a spectrum
of possibilities for dynamic activity that may be employed by real
dendrites.

The first analytical study of nonlinear interactions between
spikes at the branching points of dendritic trees was published in
1969 by Pastushenko, Markin, and Chizmadzhev, and the geometry
of the simple bifurcation they considered is sketched in Fig. 6-13.
Basically they were interested in finding the conditions under
which an action potential incoming on one or both daughters would
induce an outgoing pulse on the parent fiber.  If both $A$ and $B$
on daughters ① and ② are required to induce a pulse $C$ on the
parent, the branch can be described by the Boolean equation

$$C = A \ (AND) \ B \qquad\qquad (6.6.1)$$

If either pulse A or pulse B can induce a pulse C, we can write

$$C = A \ (OR) \ B \qquad (6.6.2)$$

while if neither A nor B can induce a pulse C

$$C = 0 \qquad (6.6.3)$$

In these Boolean (or logic) equations, the variables (A, B, and C) can assume only the values 0 and 1. The corresponding arithmetic is:

| | |
|---|---|
| 0 (AND) 0 = 0 | 0 (OR) 0 = 0 |
| 0 (AND) 1 = 0 | 0 (OR) 1 = 1 |
| 1 (AND) 0 = 0 | 1 (OR) 0 = 1 |
| 1 (AND) 1 = 1 | 1 (OR) 1 = 1 |

A nonlinear diffusion equation (2.30) was used to describe each of the three fibers (Pastushenko, Markin, and Chizmadzhev, 1969a, b)

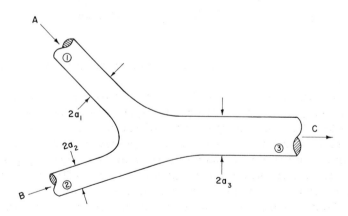

FIGURE 6-13.   General geometry for a branching dendrite.

$$\frac{\partial^2 v_k}{\partial x_k^2} - r_{sk} c_k \frac{\partial v_k}{\partial t} = r_{sk} j_{ik}, \qquad k = 1, 2, 3 \qquad (6.6.4)$$

where the three distance coordinates $(x_1, x_2, x_3)$ increase away from the junction. Using (2.21a), Kirchhoff's current law at the junction becomes

$$\frac{1}{r_{s1}} \frac{\partial v_1}{\partial x_1} + \frac{1}{r_{s2}} \frac{\partial v_2}{\partial x_2} + \frac{1}{r_{s3}} \frac{\partial v_3}{\partial x_3} = 0 \qquad (6.6.5)$$

Pastushenko and coworkers used the Markin-Chizmadzhev model (see Section 4-4) to describe the incoming pulses, demonstrating once again the remarkable usefulness of that simple description. As is indicated in (4.4.1), the membrane ion currents ($j_{i1}, j_{i2}$ and $j_{i3}$) remain zero until the membrane voltage passes a threshold level $V_1$ after which they jump to $J_{1k}$ for a time $\tau_{1k}$, then jump to a level $J_{2k}$ for a time $\tau_{2k}$, then return to zero (see Fig. 4-15). The voltage pulse returns to zero if the net ionic charge transfer is zero which requires

$$J_{1k} \tau_{1k} = J_{2k} \tau_{2k}, \qquad k = 1, 2, 3 \qquad (6.6.6)$$

As a further simplification, they assume

$$\frac{J_{11}}{c_1} = \frac{J_{12}}{c_2} \quad \text{and} \quad \frac{J_{21}}{c_1} = \frac{J_{22}}{c_2}$$

which is to be expected if both daughters have identical membranes.

The condition for an action potential to appear on the parent fiber is simply that the voltage $v_3(0, t)$ must exceed the threshold value $V_1$. Thus only a linear diffusion calculation of the sort outlined in Section 5-3 is required. However, the admittance matching conditions (geometric ratio equals unity) is _not_ satisfied for the incoming pulses on the daughters, so reflections by the junction from each of the daughter pulses will affect both daughters. Pastushenko, Markin, and Chizmadzhev account for this by defining pulse velocities $u_1$ and $u_2$ for those points on the incoming waves that first reach the threshold voltage $V_1$.    The coordinates $L_1$ and $L_2$ defined by

$$v_1(L_1, t) = V_1 = v_2(L_2, t) \qquad (6.6.7)$$

move as

$$L_1 = -u_1 t$$
$$\qquad\qquad (6.6.8a, b)$$
$$L_2 = -u_2(t - \tau)$$

where $\tau$ is a time delay between the incoming pulses.    It is important to note that the velocities $u_1$ and $u_2$ are only constant when the pulses are far from the junction $(L_1, L_2 \rightarrow -\infty)$.    As they approach the junction, the above mentioned reflections will change the velocities by influencing the condition (6.6.7).

As a first problem, Pastushenko and Markin consider only a single pulse coming in (say) on daughter ①.    Neglecting external series resistance, the junction has a geometric ratio

$$GR = \frac{a_2^{3/2} + a_3^{3/2}}{a_1^{3/2}} \qquad (6.6.9)$$

From the results of Section 4-6 on propagation through a discontinuity, we should expect a pulse to form on the parent and the other daughter whenever this $GR < 1$.    For $GR > 1$ and sufficiently large, on the other hand, blocking of conduction should occur. In agreement with (4.6.34), Pastushenko and coworkers find the condition for a pulse on the parent to be

$$\frac{a_2^{3/2} + a_3^{3/2}}{a_1^{3/2}} < \kappa + 1.11 \kappa^{\frac{1}{2}} - 1.69 \qquad (6.6.10)$$

Following the ideas presented in Section 6-3 we can use the condition of approximate conservation for leading edge charge to obtain a similar expression.    From (6.3.9) the incoming charge carries on the leading edge of pulse A (daughter ①), is $\zeta k a_1^{3/2}$

where $\zeta$ indicates the factor by which this charge increases as the pulse slows down [ see (6. 3. 10)]. This incoming charge will divide between the parent and the other daughter in a ratio determined by the corresponding characteristic admittances [see (6. 3. 7)]. Thus the parent receives a fraction $[a_3^{3/2}/(a_2^{3/2} + a_3^{3/2})]$ of the input charge, and it requires a charge $\alpha k a_2^{3/2}$ to achieve threshold [ see (5. 3. 24)]. Then the condition for a pulse on the parent is (Scott, 1973b)

$$\zeta k a_1^{3/2} \frac{a_3^{3/2}}{a_2^{3/2} + a_3^{3/2}} > \alpha k a_3^{3/2}$$

or

$$\frac{a_2^{3/2} + a_3^{3/2}}{a_1^{3/2}} < \frac{\zeta}{\alpha} \tag{6.6.11}$$

The relation between (6. 6. 11) and (6. 6. 10) is exactly the same as that found between (5. 3. 25) and (4. 6. 34). Following the notation of Pastushenko and coworkers we can denote the right-hand side by $K$. Thus

$$K \equiv \kappa + 1.11 \kappa^{\frac{1}{2}} - 1.69 \quad \text{or} \quad \frac{\zeta}{\alpha} \tag{6.6.12}$$

depending on which theoretical point of view is being assumed. For fibers which correspond to the Hodgkin-Huxley axon, blocking occurs for

$$GR \quad (5.5)^{3/2} = K$$

Suppose now that both daughter fibers have the same radius $(a_1 = a_2)$ in Fig. 6-13. Then (6. 6. 10) and (6. 6. 11) indicate that the condition to avoid block of a single input pulse is

$$\frac{a_3}{a_1} < (K - 1)^{2/3} \tag{6.6.13}$$

If this condition is satisfied, the branch functions as an "OR" junction defined by (6. 6. 2). If it is not satisfied, the condition for an AND" junction, from (5. 3. 25), is

$$(K - 1)^{2/3} < \frac{a_3}{a_1} < (2 K)^{2/3} \tag{6.6.14}$$

The "AND" condition is defined in (6. 6. 1). If the ratio $(a_3/a_1)$ is too large to satisfy (6. 6. 14), then the junction is represented by (6. 6. 3).

Consider next the "tufted" branching indicated in Fig. 6-14 that Ramon-Moliner (1962) described as typical for dendrites of sensory neurons. We assume that n daughters of equal radius $a_1$ branch from a single parent with radius $a_2$, and ask what is the <u>threshold</u> <u>number</u>, $\theta$, of daughters that must be simultaneously active in order to induce an action potential on the parent. Again the concept of approximately conserved leading edge charge is convenient to apply. The incoming charge on $\theta$ daughters is $\theta \zeta k a_1^{3/2}$, and the fraction of this charge that enters the parent is $a^{3/2}/[(n-\theta)a_1^{3/2} + a_2^{3/2}]$. The charge required to bring the parent to threshold is $\alpha k a_2^{3/2}$. Thus $\theta$ is determined by the equation

$$\theta \zeta k a_1^{3/2} \frac{a_2^{3/2}}{(n - \theta)a_1^{3/2} + a_2^{3/2}} = \alpha k a_2^{3/2}$$

which implies

$$\theta = \frac{\left(\dfrac{a_2}{a_1}\right)^{3/2} + n}{K + 1} \tag{6.6.15}$$

where, as in (6. 6. 12), $K \equiv \zeta/\alpha$.

An important advantage of the analysis by Pastushenko, Markin, and Chizmadzhev over the concept of approximate charge conservation is that it permits study of nonsynchronous pulse inputs. For $0 < \tau < \infty$ in (6. 6. 8b) they show (1969a) that an effect of pulse interaction is to bring the two pulses into closer synchronism since the second pulse to arrive is decelerated less than

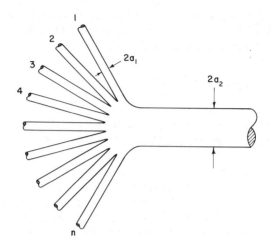

FIGURE 6-14.   Geometry of a tufted branching dendrite.

the first.   In a later work (1969b) they study the voltage $v(0, t)$
which appears at the input to the parent fiber and show that it has
two maxima for sufficiently large values of $\tau$.   Defining $\tilde{t}$ as
the value of $t$ at which

$$\frac{dv(0, t)}{dt}\bigg|_{t = \tilde{t}} = 0 \quad \text{and} \quad \frac{d^2 v(0, t)}{dt^2}\bigg|_{t = \tilde{t}} < 0$$

the situation is qualitatively as shown in Fig. 6-15.   For $\tau < \tau_v$
there is only one peak of voltage presented to the parent fiber.
For $\tau > \tau_v$ there are two maxima indicated by two branches in
$\tilde{t}(\tau)$.   The larger maximum is shown as a solid line and the smaller
by a dashed line.   At $\tau = \tau_0$, the larger maximum becomes smaller and the smaller becomes larger.   This is indicated by a discontinuity in the solid line on Fig. 6-15.   Thus Pastushenko and coworkers suggest the possibility of a rather sensitive control of
delay operating along the following lines.   Suppose the parent
fiber diameter is adjusted so it will just fire at the first maximum
$(\tilde{t} = \tilde{t}_1)$ when $\tau = \tau_0$.   Then a slight inhibition in the vicinity of
the junction (at $t = \tilde{t}_1$) could cause the parent not to fire on the
first maximum but fire instead on the second.   The "slight inhibition" would then be able to introduce a signal delay equal to
$\tilde{t}_2 - \tilde{t}_1$ (see Fig. 6-15).

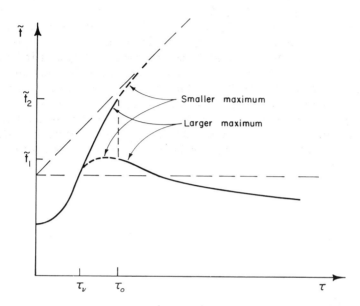

FIGURE 6-15.  Diagram related to the calculation of nonlinear
pulse interaction at a dendritic branch by
Pastushenko, Markin, and Chizmadzhev (see
text for details).

   Although it is certainly of great value to have the relatively
simple analytical results developed by Pastushenko and coworkers,
we must remember that the Markin-Chizmadzhev model for nerve
pulse propagation is an approximation.   Thus the numerical studies by Berkinblit, Vvedenskaya, Gnedenko, et al. (1971) of a
branching (as in Fig. 6-13) where the fibers are represented by
the Hodgkin-Huxley equations are also most interesting.   The
first case they consider is the effect of nonsynchronous pulse inputs on the daughter fibers.   They assume radii ratios $a_1 : a_2 : a_3 =$
$1 : 1 : 5.5$.   From the results discussed in Section 4-6 we can expect that a single pulse should not fire but both together should.
This is essentially the condition for an "AND" junction given by
(6. 6. 14).   As shown in Fig. 6-16, Berkinblit and coworkers have
computed the voltage at the junction, $v(0, t)$, for various values
of time delay, $\tau$, between the incoming pulses.   Evidently synchronism of the input pulses to within about one msec is required

to insure performance as an "AND" junction.   For a change in delay of the input pulses from 0. 9 to 0. 95 msec, the delay of the
output pulse is increased by about a millisecond.   This seems to
confirm the discontinuity in output delay around $\tau = \tau_o$ predicted
by Pastushenko, Markin, and Chizmadzhev, (1969b) and indicated
on Fig. 6-15.   The "space-time" trajectories for the voltage maxima are shown in Fig. 6-17.   In Fig. 6-17a the time delay between
input pulses is zero, whereas in Fig. 6-17b, $\tau = 0. 8$ msec.   The
leading pulse is slowed much more in the second case confirming
the prediction by Pastushenko, Markin, and Chizmadzhev, (1969a)
that such an interaction tends to bring the pulse peaks into closer
synchronism.

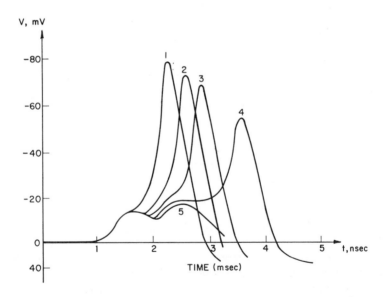

FIGURE 6-16.    Plot of voltage against time at a dendritic branch
with ratios $a_1 : a_2 : a_3 = 1 : 1 : 5$ described by the
Hodgkin-Huxley equations.   Delay between incoming pulses (in msec): (1), 0. 6; (2), 0. 8;
(3), 0. 9; (4), 0. 95; and (5), 1. 0 (redrawn from
Berkinblit, Vvedenskaya, Gnedenko, et al., 1971).

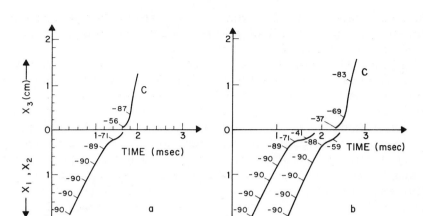

FIGURE 6-17.    Space-time plots of pulse maxima for the calculation of Fig. 6-16, where distances are measured
from the branch point: (a) synchronized input
pulses; (b) pulse input on daughter B delayed by
0. 8 msec (redrawn from Berkinblit, Vvedenskaya,
Gnedenko, et al. , 1971).

The second example discussed by Berkinblit and coworkers is
for $a_1 : a_2 : a_3 = 1 : 1. 5 : 5$.    In this case a pulse on daughter ①
alone is unable to induce an action potential on the parent while
a pulse on daughter ② alone can.    In Fig. 6-18 is shown the
junction potential for varying degrees of delay of pulse B (on
② ) behind pulse A (on ① ).    Here it is most interesting to
note that for

$$1. 7 \text{ msec} < \tau < 2. 8 \text{ msec} \qquad (6. 6. 16)$$

the pulse A <u>inhibits</u> the formation of a pulse on the parent.    When
$\tau$ lies within this range, the Boolean character of the junction can
be expressed by

$$C = B(AND) \ NOT(A) \qquad (6. 6. 17)$$

where the definition of the Boolean function NOT(·) is

$$NOT(0) = 1 \quad \text{and} \quad NOT(1) = 0 \qquad (6. 6. 18)$$

Thus have Berkinblit and coworkers demonstrated how a simple bi-furcation using a physiologically reasonable (Hodgkin-Huxley) description of the fibers can perform the logical operation of <u>inhibition</u>.    Corresponding space-time plots for trajectories of the pulse maxima are presented in Fig. 6-19 for values of $\tau$ that are too short (a) and too long (b) to satisfy the inhibit condition (6. 6. 16).    In these cases a pulse  B  on  ②  induces a pulse  A on the other daughter  ①  as well as on the parent.    Berkinblit and coworkers indicate how this effect might be applied to make a "ring oscillator" and a "motion detector".    Finally, they consider a simple bifurcation with the radii ratios  $a_1 : a_2 : a_3 = 1:1:5.$ Here each input alone can excite a pulse on the parent so the Boolean "OR" function expressed in (6. 6. 2) is realized.    For $\tau <$ 1. 4 msec the incoming pulses are essentially synchronized and a pulse travels outward only on the parent fiber.    For $\tau > 1.5$ msec the pulse  A  on  ①  induces an outward pulse on the parent and also outward on daughter  ② .    This outward pulse destroys by collision (see Section 6-1) the incoming pulse  B  on  ② .

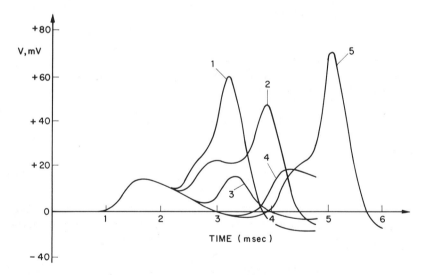

FIGURE 6-18.    Plot of voltage against time at a Hodgkin-Huxley dendritic branch with ratios  $a_1 : a_2 : a_3 = 1:1.5:5.$ Delay between incoming pulses (in msec): (1), 1. 5; (2), 1. 6; (3), 2; (4), 2. 8; (5), 3 (redrawn from Berkinblit, Dudzyavichus, Chailakhyan, et al., 1971).

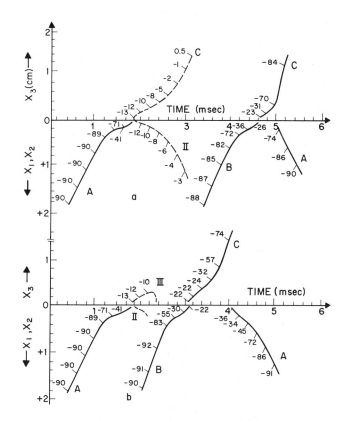

Figure 6-19.   Space time plots of pulse maxima for the calculations
of Fig. 6-18.   The pulse delay in (a) is too long
and in (b) too short to satisfy the "inhibit" condition.


It should be clear that the possibilities for information processing in active dendrites have not been exhausted; indeed the
study has scarcely begun.   But it is important to note that some
physiological confirmation of these effects is also available.
There are several studies which indicate that action potentials do
propagate at least on the larger branches of some dendritic trees
(Lorente de Nó, 1947; Cragg and Hamlyn, 1955; Eyzaquirre and
Kuffler, 1955; Fatt, 1957; Spencer and Kandel, 1961; Hild and
Tasaki, 1962; Anderson, Holmquist, and Voorhoeve, 1966;
Luk'yanov, 1970; Korn and Bennett, 1971, 1972) and, in particular,

those of cerebellar Purkinje cells (see Fig. 6-1) (Llinás, Nicholson, Freeman, and Hillman, 1968; Nicholson and Llinás, 1971; Llinás and Nicholson, 1971). In addition, Tauc and Hughs (1963) have demonstrated both "OR" and "AND" operation of branches during antidromic (backward) stimulation of mollusc axons. The sketch of these experiments is indicated in Figs. 6-20 and 6-21. Inputs are presented to two branching axons and the output is recorded through a microelectrode in the cell body. From previous studies by Tauc (1962a, b) it was possible to distinguish between the large output pulse from firing of the cell body and the smaller pulse associated with firing of the axon. It was not difficult to fire the axon but not the cell body. The "OR" function (6.6.2) is demonstrated in Fig. 6-20. Stimulation of either branch results in firing of the axon. The "AND" function (6.6.1) is demonstrated in Fig. 6-21. Stimulation of both branches with pulses sufficiently synchronized (see Fig. 6-16) is necessary to fire the axon. Tauc and Hughs conclude with the observation: "It follows from this study that a molluscan nerve cell may assume the functions of several neurons if these are considered in their classical sense as units of nervous activity. "

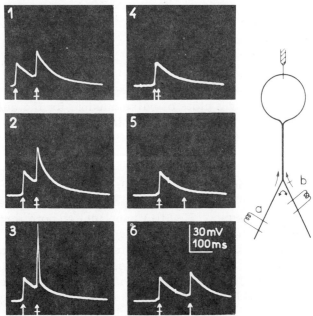

FIGURE 6-20.    Demonstration of the "OR" function at a branching mollusc axon by Tauc and Hughs (1963). [The large peak in (3) is caused by firing of the cell body. ]

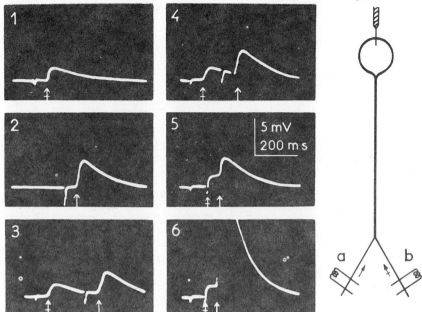

FIGURE 6-21.   Demonstration of the "AND" function at a branching mollusc axon by Tauc and Hughs (1963).

## 7.   PULSE-TRAIN DYNAMICS

In this chapter we are primarily concerned with the interaction
of individual pulses.   This is essentially a simplifying assumption
since normal neural activity involves the propagation of pulse
trains.   The present author anticipates that the dynamics of pulse
trains will become of increasing interest in the near future.   As
an introduction to such studies, consider the propagation of compressed regions on a nerve pulse train that is approximately periodic.

Whitham (1974) has developed a technique for finding solutions
to nonlinear wave problems that are almost periodic, but for which
the frequency, f, wavelength, $\lambda$ , and amplitude, A, are slowly
varying functions of space and time.   Such almost periodic solutions are not sinusoidal (often they are elliptic functions), and
the corresponding dispersion equation is of the form  $f = F(\lambda, A)$.
Two quasilinear equations for the slow evolution of $f, \lambda$ , and  A
are obtained from variation of a Lagrangian density that has been
averaged over a cycle of the periodic wave.   Such a Lagrangian
density can be obtained from an energy-conservation law (4.4).   A
third equation is conservation of wave crests

$$\frac{\partial(\frac{1}{\lambda})}{\partial t} + \frac{\partial f}{\partial x} = 0 \qquad (6.7.1)$$

For nerve fiber problems, we do not have conservation of energy; propagation is governed instead by the power balance condition (4. 3).   Furthermore, as the results obtained by Rinzel and Keller indicate (see Fig. 4-14), the frequency, amplitude and wavelength for a stable periodic wave are fixed by the local propagation velocity

$$u = f\lambda \qquad (6.7.2)$$

Thus $f = f(u)$, $\lambda = \lambda(u)$, and $A = A(u)$ so only (6. 7. 1) is needed to describe the slow evolution of $f, \lambda$, and A.   Conservation of wave crests becomes

$$\frac{\partial u}{\partial t} + U(u)\frac{\partial u}{\partial x} = 0 \qquad (6.7.3)$$

where

$$U(u) \equiv \frac{df}{d(\frac{1}{\lambda})} \qquad (6.7.4)$$

is a <u>nonlinear</u> <u>group</u> <u>velocity</u>.   For the periodic waves in Fig. 4-14, typical plots of $f$ against $1/\lambda$ were presented in Fig. 5-1.   It is interesting to note that the boundary for spatial instability found by Rinzel (5. 1. 46) is simply

$$U = 0 \qquad (6.7.5)$$

Along the stable (high velocity) branch it is clear that

$$U(u) < u \qquad (6.7.6)$$

as was noted by Rinzel and Keller (1973).   Thus (see Fig. 6-22) a compressed region in a pulse burst should drift to the rear.   This is because pulses are arriving at the rear with a speed greater than that of the compressed region.   Eventually (6. 7. 3) predicts the onset of rear end "shocks" which must, of course, be interpreted as an indication that the primary assumption of a slowly

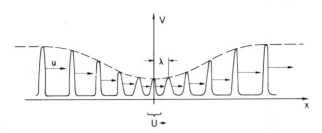

FIGURE 6-22.   A slowly varying train of nerve pulses.

varying wave train is no longer valid.   But the question of "rear
end collisions" (Crane, 1964) may be important for a nerve fiber
just as it is in a corresponding study of automobile traffic dynamics (Whitham, 1974).

Some progress toward the understanding of such effects has
been reported by Donati and Kunov (1976) in connection with their
experimental study of double pulse propagation along a squid giant
axon.   To predict the ratio of the velocity of the second pulse to
that of the first (leading) pulse, they calculated the change in
membrane conductance in the wake of the first pulse due to slow
potassium turn-off.   This change in effective resting potential was
introduced into (4.2.17) in order to calculate the change in velocity of the second pulse.   Substantial agreement was obtained between predicted and observed velocity ratios for the two pulses.
For a particular axon they found a "locking" effect at a spacing of
about 15 cm (7.5 msec).   If the pulse spacing was between 15 cm
and 20 cm, the second would travel faster than the first, but with
closer spacing than 15 cm the second pulse would propagate more
slowly.

Telesnin (1969) has studied the stability of an arbitrary number of pulses propagating on ring (or loop) of active fiber.   Assuming an enhancement to follow the refractory phase of each pulse,
he shows that the pulses will be equally spaced for a short ring
but should travel in a single compact group for a ring which is
sufficiently long.

## 8.   THE McCULLOCH-PITTS NEURON

In 1943 McCulloch and Pitts proposed that neurons might be approximately described by the following physical assumptions:

1.    The activity of the neuron is an "all-or-none" process.

2.    A certain number of synapses must be excited within the period of latent addition in order to excite a neuron at any time.

3.    The only significant delay within the nervous system is synaptic delay ($\tau$).

Under these assumptions McCulloch and Pitts began the development of a calculus of neural nets [see McCulloch (1965) for a carefully edited collection of the important papers]. Analytical implementation of the ideas expressed in assumptions 1-3 can be performed in several ways.    Under assumption 1 it is inviting to represent the activity of a neuron as a logical proposition and write the neuron output as

$$\Psi(t + \tau) = \sigma \left[ \sum_i \alpha_i \phi_i(t) - \Theta(t) \right] \tag{6.8.1}$$

where

$$\sigma[x] = +1 \quad \text{for} \quad x > 0$$
$$= -1 \quad \text{for} \quad x \leq 0 \tag{6.8.2}$$

is the signum function.    In (6.8.1) $\Psi$ takes values of $\pm 1$, whereas the synaptic weights ($\alpha_i$) and the threshold ($\Theta$) are real numbers.    The "synaptic delay" is represented by $\tau$ and often it is convenient to assume time to be quantized in units of $\tau$.    Equation (6.8.1) fails to express the possibility that a firing decision might depend on inputs at times more remote than $\tau$ (dendritic memory), but this can be included by writing

$$\Psi(t) = \sigma \left[ \sum_i \int_{-\infty}^{t} w_i(t - t') \phi_i(t') dt' - \Theta(t) \right] \tag{6.8.3}$$

where $w_i(t)$ is a (real) function that: (a) is zero for negative argument, (b) rises to a maximum in a time of order $\tau$, and (c) eventually falls back to zero as $t \to \infty$. On a longer time scale it is often of interest to consider the neuron as an analog processor of information expressed as the rates of pulse trains. Then we can write

$$F(t) = S[\sum_i \alpha_i f_i(t)] \qquad\qquad (6.8.4)$$

Here $F$ and the $f_i$ are real positive functions of time representing average firing rates and $S[\cdot]$ is a "sigmoid" type monotone increasing function.

In their initial study, McCulloch and Pitts (1943) emphasized the distinction between neural nets with "circles" (i. e., closed causal paths for logical feedback) and those without. "Nets without circles" lack the capacity for reverberatory activity and thus are analyzed with much less difficulty. In the survey of approaches to neural net analysis presented in Chapter 7, nets without circles are considered first. But (6.8.1), (6.8.3), and (6.8.4) are approximate representations of assumptions 1-3, which, in turn, seem an outrageously oversimplified description of the multiplex neuron in Fig. 6-3. One justification is to suppose a single real neuron to be represented by several hundred or more formal neurons. This net of formal neurons to describe a real neuron may be without circles; but counting is affected and the widely quoted number of $10^{10}$ cortical neurons could perhaps be interpreted as $10^{13}$ "formal neurons."

Since the objective in Chapters 7 and 8 is to suggest some difficulties in establishing a "calculus of mind," the formal neuron of McCulloch and Pitts is accepted as an appropriately conservative assumption.

# 7

# Neural Networks

i cry no quarter of my age and call
on coming wits to prove the truth
of my stark venture into fates cold hall
where thoughts at hazard cast the die for sooth

Warren S. McCulloch

## 1. NETS WITHOUT CIRCLES

The study of neural nets without closed causal loops was
vigorously pursued from the late 1950s to the mid 1960s. One
focus of this activity was the "Perceptron" idea that was begun
at Cornell University as an attempt to construct computing machines that mimic brains (Rosenblatt, 1958, 1962; Block, 1962;
Block, Knight, and Rosenblatt, 1962). A second was the ADALINE
(acronym for ADAptive LINear Element) developed at Stanford University to implement some of the ideas on reliability of computation that had concerned von Neuman (Widrow and Angell, 1962;
Nagano, Ohteru, and Kato, 1967) and the related Lernmatrix introduced by Steinbuch (1961). An extensive bibliography of this work
is included in the excellent book Learning Machines (1965) by
Nilsson (of which only the most elementary concepts are cited
here).

Suppose the $\phi_i$ values in (6.8.1) are coordinates in an n-
dimensional pattern space and $\Psi$ is to indicate which regions
satisfy a certain condition ($\Psi = +1$) and which do not ($\Psi = -1$).
Defining a weight vector

$$\overline{W} \equiv (\alpha_1, \alpha_2, \cdots, \alpha_n, \theta) \tag{7.1.1}$$

and an <u>augmented</u> <u>pattern</u> <u>vector</u>

$$\overline{P} \equiv (\phi_1, \phi_2, \cdots, \phi_n, -1) \tag{7.1.2}$$

the threshold condition (at which the argument of $\sigma[\cdot]$ is zero) is

$$\overline{W} \cdot \overline{P} = 0 \tag{7.1.3}$$

This is evidently an $(n-1)$ dimensional hyperplane in the space of the $\phi_i$ values that attempts to divide those regions for which $\Psi$ should be $+1$ from those where $\Psi$ should be $-1$. As a simple example consider Fig. 7-la, where $n = 2$ and $\Psi_A$ is the predicate

$$\Psi_A = +1 \quad \text{if the pattern has property } A$$

$$= -1 \quad \text{if the pattern does not}$$

In this case there is a one-dimensional hyperplane $(\alpha_1\phi_1 + \alpha_2\phi_2 = \theta)$ that discriminates the patterns with property $A$. A slightly more complex situation is indicated in Fig. 7-1b, where a single hyperplane that can discriminate property $A$ does not exist. However, using the two predicates $\Psi_1$ and $\Psi_2$ indicated on the figure, it is clear that $\Psi_A$ can be logically computed as

$$\Psi_A = \Psi_1 \ (\text{AND}) \ \Psi_2 \tag{7.1.4}$$

Often it is interesting to "train" (7.1.1) to classify correctly. Let us suppose it does not. That is to say there is some augmented pattern $\overline{P}_1$ with property $A$ for which $\Psi = -1$ or $\overline{W} \cdot \overline{P} < 0$. Changing the weight vector to

$$\overline{W} \to \overline{W}' = \overline{W} + c\overline{P}_1 \tag{7.1.5}$$

means that

$$\overline{W}' \cdot \overline{P}_1 = \overline{W} \cdot \overline{P}_1 + c|\overline{P}_1|^2$$

which is greater than zero if $c$ satisfies the inequality

$$c > - \frac{\overline{W} \cdot \overline{P}_1}{|\overline{P}_1|^2} \tag{7.1.6}$$

Note that the threshold condition $\overline{W} \cdot \overline{P}_1 = 0$ is a hyperplane perpendicular to $\overline{P}_1$ in the space of the weight $\overline{W}$. Thus a change of $\overline{W}$ in the direction of $\overline{P}_1$ is the shortest distance to cross the threshold and, in this sense, (7.1.5) is an efficient scheme for weight adjustment. The <u>fundamental</u> <u>training theorem</u> for a Per- is essentially that, assuming a discriminating hyperplane to exist, one will be found by iterating the weight adjustment indicated in (7.1.5). A satisfactory hyperplane is, of course, not necessarily unique.

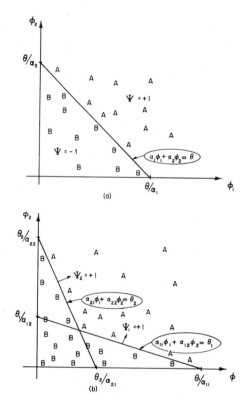

FIGURE 7-1.  Discriminant hyperplanes for pattern recognition by Perceptrons.

These ideas are so simple and appealing that it is difficult to avoid becoming overly enthusiastic. Thus the careful evaluation of the Perceptron concept by Minsky and Papert (1969) is an important contribution which should be carefully studied by all who are interested in this subject. Choice of the $\phi$ functions is a crucial consideration, so Minsky and Papert restrict themselves to the situation sketched in Fig. 7-2. Patterns are presented to a "retina" of R points each of which may be black or white, and each $\phi$ is computed from a certain subset of the retinal points. A Perceptron is defined as follows. Consider a pattern, X, a property or predicate $\Psi$ (e. g. , convexity, connectedness, the letter A, etc. ), and a family of functions

$$\Phi = \{\phi_1, \phi_2, \cdots, \phi_n\} \tag{7.1.7}$$

The predicate $\Psi(X)$ is <u>linear</u> with respect to $\Phi$ if there exist real numbers $\alpha_1, \dots, \alpha_n$ and $\theta$ such that $\Psi(X)$ is "true" (i. e. , +1) if and only if

$$\alpha_1\phi_1(X) + \cdots + \alpha_n\phi_n(X) > \theta \tag{7.1.8}$$

We can then write

$$\Psi(X) = \sigma\left[\sum_i \alpha_i\phi_i(X) - \theta\right] \tag{7.1.9}$$

which constitutes a Perceptron for the predicate $\Psi(X)$. If pattern X has the property $\Psi$, $\Psi(X) = +1$; if not, $\Psi(X) = -1$.

The repertory of a set $\Phi$, $L(\Phi)$, is defined as the set of all predicates that can be computed as in (7.1.9) with appropriate choice of the weights. The "game" of Minsky and Papert is to put restrictions on $\Phi$ to see what can be learned about the corresponding $L(\Phi)$. They are particularly concerned with the following problems:

1.  Considering the retina as an R-dimensional space of $2^R$ pattern vectors obscures the <u>real</u> geometrical properties of patterns on R. There are meaningful geometric properties that cannot be computed by Perceptrons.

2.  Size range for the weights is an important consideration. Predicates in $L(\Phi)$ requiring an impractically large range of the $\alpha$ values are of limited interest.

3.    Time of convergence must be considered in evaluating the
learning scheme.  A Perceptron can always "learn" by
cycling through all weight vectors, but this is impractical.

   Minsky and Papert define <u>diameter</u> <u>limited</u> Perceptrons for
which the inputs to any  φ function lie within a fixed distance and
<u>order</u> <u>limited</u> Perceptrons for which the number of points in the
retina seen by each φ function is bounded. Typical negative results
obtained for computation of "connectedness" are  (a)  no diameter-
limited Perceptron can compute connectedness and (b)  an order
limited Perceptron must be of order  R (i. e. ,  some  φ function must
look at every point in the retina) to compute connectedness.  Typical positive results are that simple geometrical figures (triangles,
rectangles,  alphabetical letters,  etc. ) can be computed by diameter limited Perceptrons and by Perceptrons limited to low order.

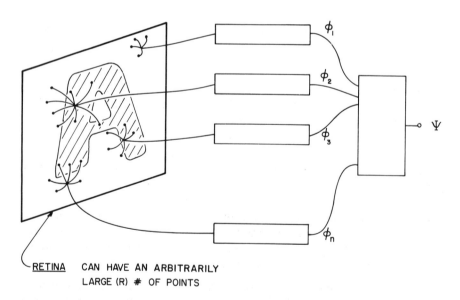

<u>RETINA</u>   CAN HAVE AN ARBITRARILY
         LARGE (R) # OF POINTS

FIGURE 7-2.   Diagram of a Perceptron.

Perceptron ideas have been used to train a digital computer to play checkers (Samuel, 1959), and recently it has been suggested independently by Marr (1969) and Albus (1971) that the Purkinje cell of the cerebellum (see Fig. 6-1) may be essentially a Perceptron that mediates muscular activity.  However, one should not jump to conclusions.  Considerably greater computing power would be obtained (see Fig. 7-1b) if logical decisions were made at dendritic branches (see Fig. 6-3).  Marr (1970) has extended his picture of the cerebellum to a detailed discussion of information processing in the cerebral cortex.  Here, however, it is necessary to consider also the influence of reverberatory activity.

## 2.  REVERBERATORY NEURAL NETS

In 1949 Hebb published his classic Organization of Behavior in which he attempted to bridge the gap between neurophysiology and psychology by postulating the existence of a new hierarchical entity that he termed the cell assembly.  A carefully developed introduction to this concept is contained in his textbook (Hebb, 1972) but the central notion is that:

> Any frequently repeated, particular stimulation will lead to the slow development of a "cell assembly, " ..., capable of acting briefly as a closed system, delivering facilitation to other such systems and usually having a specific motor facilitation.  A series of such events constitutes a "phase sequence" - - the thought process.  Each assembly action may be aroused by a preceding assembly, by a sensory event, or - - normally - - by both.  The central facilitation from one of these activities on the next is the prototype of "attention".

Hebb, a psychologist himself, marshaled substantial psychological evidence to support this view against the conflicting claims of "field theory, " which denied the importance of individual neural connections and "switchboard theory, " which, in turn, asserted direct connections between sensory and motor neurons throughout the cortex.  As the electrically induced experiential responses described in Chapter 1 indicate, something like the phase sequence is a palpable fact of thought.  It can also be experienced as a "jumping" between perceptions of profile and front views in Picasso's 1937 painting of Marie-Thérèse Walter (Fig. 7-3). [For a survey of the relation between art and perceptual dynamics, Arnheim's Art and Visual Perception (1954) is highly recommended. ] Another easily experienced phase sequence is the thought train of dreaming, a vivid example of which has recently been given by Roszak (1973).

FIGURE 7-3.    Picasso's "Marie-Thérèse Walter. " With practice, either mirror image can be viewed in front or profile, but practice with (b) - - the original - - doesn't help with (a).  (From Goodbye Picasso by David Douglas Duncan.  Used by permission of Grosset and Dunlap, Inc.)

A sphere appears in my dream ... and becomes a ball,
a familiar childhood toy I had forgotten.  At once the dream
begins to heap up associations around this ball.  It plays
exuberantly with the word "ball" ... with every possible
rhyme, pun, slang connotation, homonym.  Suddenly, there
are elegant people dancing on and around the ball; it has
become a fancy-dress ball ... where people are having a
ball ... balling the jack ... drinking highballs ... getting
drunk to the eyeballs .... .  Balls: a man's balls ... to
have balls ... to be on the ball.  Ball: to ball a woman ...
to ball up the job ... to bawl like a baby.  Ball: bald.
Ball: fall ... as hair falls ... leaving you bald as a
billiard ball.  People named Ball: John Ball ... George
Ball ... Lucille Ball.  Ball: Baltimore ... the Baltimore &
Ohio ... highballing down the line ... And the dream plays
too with the form of the ball, until it reflects every sort of
round, rolling, bouncing thing ... globes, planets, wheels,
balloons, bubbles, circles, eggs, oranges, coins, fireballs,
goof balls, golf balls, footballs ... a baseball which is
"the old apple" ... forbidden fruit .... .

Hebb viewed the cell assembly as a "closed solid cagework,
or three-dimensional lattice, with no regular structure, " and supposed it to develop under the following conditions.

When an axon of cell  A  is near enough to excite a cell
B  and repeatedly or persistently takes part in firing it,  some
growth process or metabolic change takes place in one or
both cells such that A's efficiency, as one of the cells
firing  B,  is increased.

Although Hebb (1949) favored a strengthening of synaptic contacts as a mechanism for learning, the above assumption is not
specific.  It does, however, consider only an increase in firing
efficiency.  In 1955 Frankel reviewed several approaches to the
realization of machines to mimic mammalian brains and suggested
a design along the lines of Hebb's theory as most promising.  Since
the difficulties of dealing analytically with such a system appeared
unmanageable, he proposed a computational investigation.  Rochester, Holland, Haibt, et al. (1956) presented the results of such a
test that modeled 99 neurons as indicated in (6. 8. 3) with no inhibitory interconnections and assumed the time to be quantized in

discrete steps of a "synaptic delay, " $\tau$. They found a diffuse
reverberation with a period of the order of the refactory time but
could not demonstrate the growth of cell assemblies.  Rochester,
Holland, Haibt, et al. (1956) then talked with Milner who was
revising Hebb's learning assumption to include inhibition as a
decreasing efficiency of firing (Milner, 1957).  They [Rochester,
Holland, Haibt, et al. (1956)] subsequently modified their program to
include inhibition in 512 neurons with six "external" neurons excited every ten time steps by organized signal plus noise.  Cell assemblies were found to develop with excitatory contacts between
cells in the same assembly and inhibitory contacts between adjacent assemblies.  Thus two such assemblies could act as a "flip-
flop" that mimics oscillation between the two perceptions in Fig.
7-3.  Griffith (1967) has demonstrated that an "habituation" effect
that gradually raises the threshold of a neuron during activity can
change such a flip-flop into the "multivibrator" indicated in Fig.
7-4.  When assembly A ("mode A" in Griffith's terminology)
fires, it inhibits assembly B, but eventually the thresholds for
neurons in assembly A rise and it becomes extinguished.  This
removes the inhibition from sssembly B permitting it to fire.  A
significant aspect of this study is the extermely rapid "turn-on"
of usually only one or two time increments.

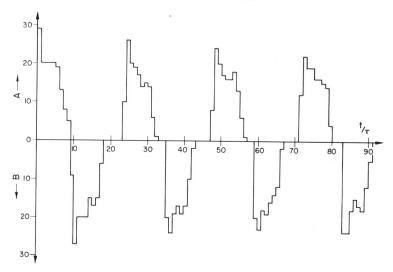

FIGURE 7-4.    Free-running "multivibrator" oscillation of two cell
assemblies with inhibitory interaction and habituation.  From A View of the Brain, by J. S. Griffith,
published by Oxford University Press, 1967.

Hebb (1949) discussed a hierarchical organization of cell assemblies, making particular reference to visual perception. Although we consider this concept in greater detail in the text that follows,it seems appropriate to introduce the gist of it here. Noting that neural mapping from the retina to area 17 of the visual cortex (see Fig. 1-4) is topological, he proposed lines and angles positioned through eye movement as the first level cell assemblies (subassemblies) localized in area 17. Since the connections from area 17 to area 18 are no longer topological but diffuse, this area can serve as a location for assemblies that organize lines and angles into perceived geometrical figures such as triangles, rectangles, circles, and so on. This notion receives some support from the studies by Hubel and Wiesel (1962) of the functional architecture of the feline visual cortex. They conclude:

> It is suggested that columns containing cells with common receptive-field axis orientations are functional units, in which cells with simple fields represent an early stage in organization, possibly receiving their afferents directly from lateral geniculate cells, and cells with complex fields are of higher order, receiving projections from a number of cells with simple fields within the same column.

In 1961 Caianiello proposed a detailed analysis of neural systmes composed of elements as in (6. 8. 3) paying particular attention to reverberatory states and their relation to thinking and consciousness. Caianiello, de Luca, and Ricciardi (1967) investigated the reverberatory character of N formal neurons described as in (6. 8. 1) with every $\theta$ (threshold) equal to zero, a _normal_ system. Time was assumed to be quantized in units of the synaptic delay, $\tau$; and a real variable, v, was introduced to represent signal strength before processing by the signum function. Thus the input to the firing decision for neuron h at time $t + \tau$ is related to the states of all neurons in the net at time t by

$$v_h(t + \tau) = \sum_k \alpha_{hk} \sigma[v_k(t)] \qquad (7.2.1)$$

Introducing a vector description of the states of the N neurons at time $t = m\tau$ as

$$\bar{v}_m \equiv \bar{v}(m\tau) \equiv \begin{bmatrix} v_1(m\tau) \\ v_2(m\tau) \\ \vdots \\ v_N(m\tau) \end{bmatrix} \qquad (7.2.2)$$

permits (7.2.1) to be written in matrix form as

$$\boxed{\bar{v}_{m+1} = A\sigma[\bar{v}_m]} \qquad (7.2.3)$$

where $A \equiv [\alpha_{hk}]$ is the $N \times N$ matrix of interconnection strengths.  A <u>reverberation</u> is defined as a sequence of states

$$\bar{v}_m \rightarrow \bar{v}_{m+1} \rightarrow \cdots \quad \bar{v}_{m+R} = \bar{v}_m \qquad (7.2.4)$$

and $R\tau$ is the <u>period</u>.  Even this greatly oversimplified model for a neural system has $2^N$ states, and a reverberation that cycled through all of these states would have a period of $2^N\tau$ sec. Taking $N = 100$ neurons and $\tau = 10^{-3}$ sec implies a period of about $10^{19}$ years, which is much longer than the age of the universe!  Caianiello, de Luca, and Ricciardi showed how the rank (K) of the connection matrix could be used to establish an upper bound on the period of a reverberation.  By partitioning the matrix A as

$$A = \begin{bmatrix} B & \vdots & B\Lambda \\ ----- & \vdots & ---- \\ MB & \vdots & MB\Lambda \end{bmatrix} \begin{matrix} \}\text{ K rows} \\ \\ \}\text{ (N - K) rows} \end{matrix} \qquad (7.2.5)$$

$$\underbrace{\text{K columns}} \quad \underbrace{\text{(N- K) columns}}$$

and defining a reduced vector $\bar{v}'_m$ as

$$\overline{v}'_m \equiv \begin{bmatrix} v_{1m} \\ v_{2m} \\ \vdots \\ v_{Km} \end{bmatrix} \quad \text{and also} \quad \overline{v}''_m \equiv \begin{bmatrix} v_{K+1,m} \\ v_{K+2,m} \\ \vdots \\ v_{N,m} \end{bmatrix} \qquad (7.2.6a,b)$$

(7.2.3) can be written

$$\begin{bmatrix} \overline{v}'_{m+1} \\ \overline{v}''_{m+1} \end{bmatrix} = \begin{bmatrix} B & \vdots & B\Lambda \\ \cdots & \vdots & \cdots \\ MB & \vdots & MB\Lambda \end{bmatrix} \begin{bmatrix} \sigma[\overline{v}'_m] \\ \sigma[\overline{v}''_m] \end{bmatrix} \qquad (7.2.7)$$

Thus

$$\overline{v}'_{m+1} = B[\sigma[\overline{v}'_m] + \Lambda\sigma[M\overline{v}'_m]] \qquad (7.2.8)$$

In other words, the reduced vector at time $m+1$ depends only on the reduced vector at time $m$. At any time $\overline{v}''_m = M\overline{v}'_m$, and clearly $\sigma[\overline{v}_m]$ has just $2^K$ states. The signum function has the property $\sigma[xy] = \sigma[x] \cdot \sigma[y]$. Thus we can write

$$\sigma[M\overline{v}'_m] = \sigma[M]\sigma[\overline{v}'_m] \qquad (7.2.9)$$

if we assume that the matrix $M$ has only a single element in each row and column. Then the period of a reverberation is bounded by

$$R \leq 2^K \qquad (7.2.10)$$

Without assuming quantized time or zero threshold, Caianello, de Luca, and Ricciardi also showed that the reverberatory system would have $(N-K)$ constants of the motion. To see this rewrite (7.2.3) as

$$\overline{v}(t) = A\sigma[\overline{v}(t-\tau)] - \overline{\theta} \qquad (7.2.11)$$

where $\overline{\theta} \equiv \text{col}(\theta_1,\dots,\theta_N)$ is a threshold vector. Taking the scalar product of (7.2.11) with a constant, $N$ component vector $\overline{\gamma}$ gives

$$g(t) = \overline{\gamma} \cdot \overline{v}(t) = \overline{\gamma} \cdot A\sigma[\overline{v}(t - \tau)] - \overline{\gamma} \cdot \overline{\theta}$$

If $\overline{\gamma}$ is chosen so

$$\overline{\gamma}A = (0, 0, \ldots, 0) \tag{7.2.12}$$

then

$$g = -\overline{\gamma} \cdot \overline{\theta} \tag{7.2.13}$$

For A an $N \times N$ matrix of rank $K$ there are $(N-K)$ independent vectors, $\overline{\gamma}$, which satisfy (7.2.12) and give $(N-K)$ constant values for $g$ in (7.2.13). See Aiello, Burattini, and Caianiello (1970) for further discussion of constants of the motion and learning invariant rank.

A system of $N$ neural elements and, therefore, $2^N$ states can also be described in terms of a state diagram that indicates the evolution that the system undergoes as time increases. For three elements there are eight states of the vector $(n_1, n_2, n_3)$ where $n_i = \pm 1$, and each of these can be designated by a point (see Fig. 7-5). Only one arrow leaves each state indicating the unique state for the next time increment, but any number of arrows can enter a state. Thus emerges a basic nerve net property, namely, irreversibility in time. Reverberations are related to the closed cycles, and "transients" to paths not included in cycles. Thus if the system is initiated in state (111) it will go through state (-111) before entering a reverberation of period 2. Kitagawa (1973) and Ishihara and Sato (1974) (see also Sato and Ishihara, 1974) have recently discussed the application of graph theory to the study of shifts and stability of reverberations.

In general each of the $N_1$ neural elements may be an arbitrary Boolean function with $N_2$ inputs. Since there are $2^{N_2}$ input combinations and the output can be $\pm 1$ for each, there are

$$2^{2^{N_2}}$$

Boolean functions of $N_2$ inputs. Each of the $N_1$ neural elements can be any of these functions so there are

$$\mathcal{N}_B = 2^{N_1 2^{N_2}} \tag{7.2.14}$$

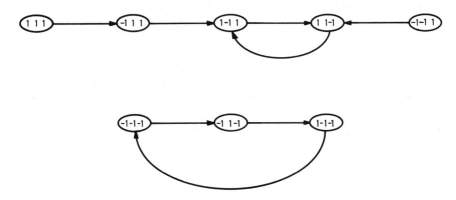

FIGURE 7-5.    One of the $2^{24}$ state diagrams for a logic machine with three elements.

possibilities for the system.   If we restrict the neural elements to the threshold calculation indicated in (6.8.1), not all of these possibilities can be realized.   The simple state diagram of Fig. 7-5, for example, demands that the first element decide $-1$ when the inputs are (111) and (-1-1-1), but must decide $+1$ for other inputs.   As was indicated in Fig. 7-1, this decision cannot be made by a threshold element.   The number of threshold logic functions of $N_2$ inputs is not exactly known, but for large $N_2$ it is approximately (Yajima, Ibaraki, and Kawano 1968)

$$2^{kN_2^2}$$

where

$$\frac{1}{2} \le k \le 1$$

The corresponding number of $N_1$ element threshold systems is then

$$\mathcal{N}_T = 2^{kN_2^2 N_1} \tag{7.2.15}$$

The ratio $(\mathcal{N}_T/\mathcal{N}_B)$ falls rapidly to zero as $N_1$ and $N_2$ increase; for $N_1 = N_2 = 7$ neural elements it is $0.548 \times 10^{-200}$.   Thus it is important to know whether a neuron computes an arbitrary Boolean function (as might be inferred from the "multiplex neuron" of Fig. 6-3) when making numerical estimates of the dynamical possibilities for a neural system.

Consider next an $N$ neuron system for which only $m \leq 2^N$ of the state transitions are specified. Yajima and coworkers define $R(m, N)$ as the ratio of threshold systems to the total number of Boolean systems with this specification and show that

$$R(\alpha N, N) \to 0 \quad \text{for} \quad \alpha > 2$$

$$\to 1 \quad \text{for} \quad \alpha < 2$$

as $N \to \infty$. Thus threshold logic imposes little restriction on the realization of an $N$ element system as long as less than $2N$ transitions in the state diagram are specified.

These considerations may be useful for extending and evaluating the discussions by Ishihara (1971a, b) of the interactions between reverberations. Referring to Fig. 7-6 he describes Pavlov's classical conditioned reflex experiment in the following manner. The food-salivation mechanism is assumed to be established as a reverberation A in area 19, where the loop of dots represents a sequence of states (as in Fig. 7-5) rather than the successive firing of individual neurons. When the bell is repeatedly rung, a corresponding train of pulses projects to areas 41 and 42 of the temporal lobe (see Fig. 1-4) eventually establishing reverberatory activity B in area 22. Simultaneous stimulation by the bell and the meat leads to the development of an interaction between assemblies A and B, eventually permitting stimulation of A by the bell via B without the presence of the meat. Ishihara has been primarily concerned with the representation of such an interaction in the context of Caianiello's neuronic equations (7.2.3). In the text that follows we consider more physical descriptions of how such a long-range interaction could occur.

The computational difficulty of dealing with a dynamic system as complex as at least one hundred formal neurons led Shimbel and Rapoport (1948) to consider the development of a statistical neurodynamics, and Rapoport (1952) used this approach to study the "ignition" phenomenon in a neural network (see also Trucco, 1952). Smith and Davidson (1962) presented a simple description of this effect which assumed a network of $N$ neurons, each having $e$ excitatory connections coming randomly from the other neurons. Time was assumed to be quantized and inputs were

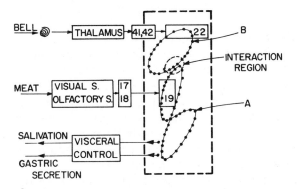

FIGURE 7-6.   Schematic representation of Pavlov's classical
conditioning experiment (redrawn from Ishihara,
1971b).

summed over  s  time units.  Calling the activity or fraction of
cells firing  F,  the probability of a cell receiving  i  units of
excitation over the previous  s  time units on  e  inputs is[*]

$$\frac{(se)!}{i!\,(se-i)!}\,F^i(1-F)^{se-i} \equiv \binom{se}{i}F^k(1-F)^{se-i}$$

The probability of firing,  S,  is the sum of this quantity for all
$i \geq \theta$  (the threshold) so

$$S(F) = \sum_{i=\theta}^{se} \binom{se}{i} F^i(1-F)^{se-i} \qquad (7.2.16)$$

An equilibrium condition for sustained activity is

$$S(F) = F \qquad (7.2.17)$$

which is easily shown to be satisfied for  F = 0  (no activity) and
F = 1  (maximum activity).  If  $1 < \theta < se$,  S(F)  has the sigmoid
shape indicated in Fig. 7-7 so there is an intermediate equilibrium
at which

---
[*] As any craps shooter should know.

$$\frac{dS}{dF} > 1 \qquad\qquad (7.2.18)$$

Thus a slight increase (decrease) in F causes a greater increase (decrease) during the next time increment and the intermediate equilibrium point is always unstable. Ashby, von Foerster, and Walker (1962) saw this effect as "something of a paradox" because the brain does, indeed, exhibit intermediate states of activity. Confusion arises here because a completely statistical representation of a nerve net fails to distinguish between (say) one percent of the neurons firing at maximum rate and all the neurons firing at one percent of the maximum rate. Thus in the context of Hebb's discussion, the paradox can be resolved by supposing that (7.2.16) and Fig. 7-7 apply to the neurons of a particular cell assembly with inhibitory connections to neighboring assemblies.

Statistical neurodynamics might be considered as a procedure for establishing the "laws of thought" from network dynamics (level 5 in Chapter 1) just as statistical thermodynamics derives the gas laws from molecular dynamics. For careful treatments see

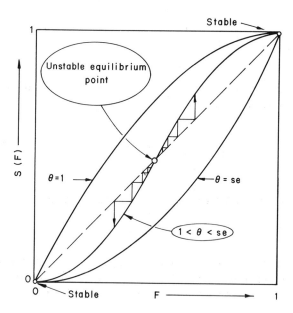

FIGURE 7-7.    Solution plot for equation (7.2.17).

Rozonoér (1969), Amari (1971, 1974), and also the combined numerical and analytic studies by Burattini and Liesis (1972).  Cowan
(1970) has discussed the possibility of establishing a Hamiltonian
formulation for neural activity, but the corresponding restrictions
on neural interconnections may be too severe.  The work of
Ventriglia (1974) is also motivated by an analogy with classical
physics.  He considers the neural system to be composed of two
types of "particles" (neurons and impulses) enclosed within a
three-dimensional space.  The neural particles are in fixed posi-
tions while the impulses form a surrounding "gas. "  Impulse-
impulse collisions are neglected, but the total number of impulses
is assumed to change after an impulse-neuron collision.  A kinetic equation is developed for evolution of the distribution function
for impulses.  The difficulty, of course, is that the organization
of a neural mass into (say) cell assemblies determines which aspects of the mass may be treated statistically, and it is just this
organization that is not evident from the study of neural nets.
Some ad hoc notion, such as the cell assembly, seems necessary
as a working hypothesis in order to proceed.

White (1961) was among the first to consider the dynamics of
cell assemblies in relation to the "ignition" phenomena displayed
by random neural nets.  For a net with $N$ neurons he began with
a simple version of the equilibrium conditions (7. 2. 17) in the form

$$F = S(P + \beta NF) \qquad (7. 2. 19)$$

where $P$ is an external input to the net and $\beta$ is a coefficient
which expresses interconnectedness.  Inhibition is included by
supposing $\beta$ to be originally negative; it gradually becomes positive as the net develops into an assembly under repeated stimulation by $P$.  Rewriting (7. 2. 19) as

$$S^{-1}(F) = P + \beta NF \qquad (7. 2. 20)$$

and referring to Figs. 7-8a, b, it is clear that for $\beta$ negative or
positive but less than the minimum slope

$$\frac{dS^{-1}(F)}{dF}\bigg|_{min}$$

there is only one firing state and $F$ will return to zero as $P$ is reduced to zero.   However, if repeated stimulation by P increases $\beta$ so it satisfies the inequality

$$\beta > \left.\frac{dS^{-1}(F)}{dF}\right|_{min} \qquad (7.2.21)$$

there can be two firing states.   Increasing P sufficiently and then reducing it again will leave the network in the upper firing state as shown in Fig. 7-8c.

For two cell assemblies, $N_1$ and $N_2$, equilibrium equations can be written

$$S^{-1}(F_1) = P_1 + \beta_{11}N_1 F_1 + \beta_{12}N_2 F_2$$

$$\qquad\qquad\qquad\qquad\qquad\qquad\qquad (7.2.22a, b)$$

$$S^{-1}(F_2) = P_2 + \beta_{21}N_1 F_1 + \beta_{22}N_2 F_2$$

Let us assume that $\beta_{11}$ and $\beta_{22}$ both satisfy the inequality (7.2.21) so the assemblies can be individually ignited.   Then if $\beta_{21}(\beta_{12})$ is also positive, it is impossible to ignite assembly 1(2) without also stimulating assembly 2(1).   This is in agreement with the computer studies by Rochester, Holland, Haibt, et al. (1956) and Griffith (1967) indicating the need for some inhibition <u>between</u> assemblies.   If $\beta_{12}$ and $\beta_{21}$ are both negative, a variety of

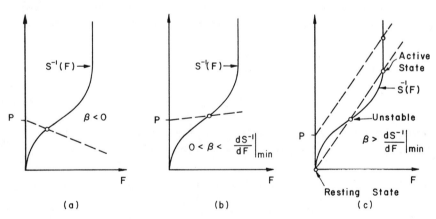

(a)            (b)            (c)

FIGURE 7-8.   Formation of a cell assembly described by equation (7.2.20): (a) mutual inhibition; (b) weak mutual excitation; (c) strong mutual excitation.

solutions to (7.2.22) can be obtained involving the ignition of one or both assemblies.

Following an initial attempt by Harth and Edgar (1967) to model the cortex as a mass of association neurons which were highly damped so reverberations would not occur, Harth, Csermely, Beek, et al. (1970) turned to the idea of cell assemblies (called netlets). They begin with a careful review of the biological evidence supporting Hebb's theory, which includes: (a) experiential response to electrical stimulation of the cortex [described in Chapter I of Penfield and Perot (1963)], (b) Mountcastle's (1957) observation of radial columns of neurons in the somatosensory cortex of the cat that appear to act as "elementary units of organization," (c) the previously cited suggestion by Hubel and Wiesel (1962) of a functional hierarchy for cell pools in the cat's visual cortex, and (d) intracortical microstimulation of the cat's motor cortex by facilitating currents as low as 2 $\mu$A (Asanuma, Stoney, and Abzug, 1968), which indicates that "the basic design of motor-sensory cortex includes radially arranged colonies of functionally related neurons." Since the cortex contains some $10^{10}$ neurons, any attempt at an holistic description must greatly reduce the number of parameters considered. Cell assemblies might serve as appropriate "macrostates" that are random in the small and organized in the large. Harth and coworkers stake out their position as follows:

It should by no means be taken for granted that such neuronal macrostates must exist, nor that their description can be made sufficiently simple to be of practical value. However both will be assumed here. It is difficult to see how significant progress in understanding the brain can ever be achieved unless these two assumptions are justified.

They then describe the ways in which cell assemblies could interact in the cortex to process information and to model the conditioned reflex. Learning is assumed to proceed according to Caianiello's (1961) adiabatic learning hypothesis, according to which the coupling coefficients [the $\alpha_{hk}$ in (7.2.1)] change slowly on the time scale for reverberatory activity. The detailed mathematical analysis is described in a companion paper by Anninos, Beek, Csermely, et al. (1970). They assume a "refractory period," r, which is related to the firing delay, $\tau$, by

$$\tau < r < 2\tau \qquad\qquad (7.2.23)$$

This means that at each time step only those neurons that fired on the previous step would be unavailable for refiring. Equation (7.2.17) then becomes modified to

$$F(t + \tau) = [1 - F(t)] S[F(t)] \qquad\qquad (7.2.24)$$

which leads to a high probability of cyclic activity between two states as in Fig. 7-4. A formulism is developed, involving macroscopic coupling coefficients, for systems containing an arbitrary number of assemblies. Numerical results from application of this formulism to a system of two assemblies is discussed by Harth, Cseremely, Beek, et al. (1970). In a series of subsequent papers, Anninos (1972a, b, 1973) and his associates [Anninos and Elul (1974), Cyrulnik, Anninos, and Marsh (1974)] present a variety of numerical results on the dynamical behavior of single-cell assemblies (see also Dunin-Barkovskii, 1970).

In a more analytical discussion, Wilson and Cowan (1972) consider two subpopulations of neurons; excitatory and inhibitory, for which the fraction firing per unit time are described by two dependent variables $E(t)$ and $I(t)$. After a cell fires, it is assumed to be refractory for a time $r$, so the fraction of excitatory cells ready to fire at time $t$ is

$$1 - \int_{t-r}^{t} E(t')dt \approx 1 - r E(t)$$

with a corresponding expression for the inhibitory cells. Assuming the delay time for firing to be a value $\tau$, Wilson and Cowan write a dynamic equation

$$E(t + \tau) = [1 - rE(t)] S_e[\beta_1 E(t) - \beta_2 I(t) + P(t)] \qquad (7.2.25)$$

where $S_e$ is a sigmoid response curve for the excitatory neurons, $P(t)$ is an external input to the excitatory neurons, and both $\beta_1$ and $\beta_2$ are positive interconnection constants. Using the approximation

$$\tau \frac{E(t + \tau) - E(T)}{\tau} \approx \tau \frac{dE}{dt}$$

(7. 2. 25) becomes an ordinary differential equation for $E(t)$

$$\tau \frac{dE}{dt} = - E + (1 - rE) S_e[\beta_1 E - \beta_2 I + P] \qquad (7. 2. 26a)$$

and similarly for $I(t)$, the inhibitory subpopulation

$$\tau \frac{dI}{dt} = - I + (1 - rI) S_i[\beta_3 E - \beta_4 I + Q] \qquad (7. 2. 26b)$$

where $Q$ is an input specifically directed to the inhibitory neurons. Equilibrium conditions from (7. 2. 26) can be written

$$\beta_2 I = \beta_1 E + P - S_e^{-1} (\frac{E}{1 - rE}) \quad \text{for} \quad \frac{dE}{dt} = 0$$

$$(7. 2. 27a,b)$$

$$\beta_3 E = \beta_4 I - Q + S_i^{-1} (\frac{I}{1 - rI}) \quad \text{for} \quad \frac{dI}{dt} = 0$$

From (7. 2. 27b) $E$ is a monotone increasing function of $I$ along the locus (in the $E - I$ phase plane) where $(dI/dt) = 0$. From (7. 2. 27a), however, $I$ is not necessarily a monotone increasing function of $E$ along the locus where $(dE/dt) = 0$. Two possibilities are displayed in Fig. 7-9, where the directions of trajectories along the equilibrium lines can be established from reference to (7. 2. 26). A smaller value of $\beta_1$ and a larger value of $\beta_4$ favor the formation of stable equilibrium points indicated in Fig. 7-9a over the oscillatory behavior in Fig. 7-9b. Transitions from one equilibrium point to the other in Fig. 7-9a can be accomplished by moving "the $(dE/dt) = 0$ locus" right and left through the excitatory input $P$ or by moving the "$(dI/dt) = 0$ locus" up and down with $Q$.

The biologically oriented reader will notice that equations (7. 2. 26) are closely related to the Volterra equations for the interaction of species. Recently Lin and Kahn (1976) have used the averaging method of Kryloff and Bogoliuboff (1947) to find conditions for existence and amplitude of limit cycles when the system is near sinusoidal oscillation.

Grossberg (1973) has investigated the interaction of inhibitory and excitatory neuron populations from a more theoretical point of view (i. e., using a "theorem-proof" format). He assumes that the interaction parameters are identical for both components so a single activity variable (let's call it $F_i$) is sufficient to describe the activity of the ith cell population, and that each population excites itself as it inhibits neighboring populations according to

$$\frac{dF_i}{dt} = -AF_i + (B - F_i)S(F_i) - F_i \sum_{k \neq i} S(F_i) + P_i \qquad (7.2.28)$$

It is interesting to compare (7.2.28) with the Wilson-Cowan equations in (7.2.26). The excitatory process, $(B - F_i)S(F_i)$, takes a form different from the inhibitory process,

$$- F_i \sum_{k \neq i} S(F_i)$$

because a recurrent "on-center, off-surround" interconnection scheme is assumed in which the cell population, $F_i$, excites only itself but inhibits neighbors within a certain halo. The multiplicative character of the inhibitory interaction is termed a "shunting" effect, which has heuristic value in assembly function as the relative levels of the inhibiting assemblies are preserved. $P_i$ is an external input to the ith assembly.

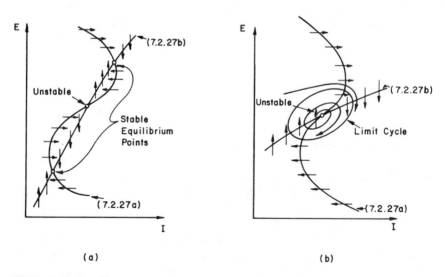

FIGURE 7-9.    Phase-plane trajectories for excitatory and inhibitory neural subpopulations from equations (7.2.26).

The objective of Grossberg's work is to analyze the dynamic properties of (7.2.28) with various assumption on the character of the function $S(\cdot)$. In particular, a sigmoid character leads to edge enhancement of patterns and to localized reverberation or short-term memory. More complex applications of this approach to processing of pattern information are discussed in Grossberg and Levine (1975), Ellias and Grossberg (1975), and in Levine and Grossberg (1976). Here influences on pattern processing of spatiotemporal parameters in excitatory and inhibitory cell populations are considered.

Having suggested the notion of a cell assembly as an atomistic entity, it is interesting to consider how many assemblies might exist in the human brain. This is a difficult question, but Legéndy (1967) has obtained some results from a very simple model. He assumes that the brain is already organized into subassemblies and discusses their organization into assemblies. The assembly and one of its subassemblies variously represents "a setting and a person who is part of it, a word and one of its letters, an object and one of its details." Interconnections are assumed evenly distributed over the brain to avoid the complications of spatial organization. Subassemblies and assemblies are like neurons in that a threshold of excitation must be exceeded for ignition, but they are also bistable. As indicated in Fig. 7-8c, they may remain in an active state (subject to the constraints of habituation) as well as in a resting state. Whereas the threshold for a subassembly is assumed to be a certain number of active neurons, the threshold for an assembly is a certain number of active subassemblies. Assembly storage has two additional advantages over neuron storage:

1. Physical damage will not destory specific assemblies, but degrade many by roughly equal degrees.

2. The growth of interconnections is much more plastic for assemblies than for neurons.

Legéndy considers the subassemblies to be formed through "weak" contacts and assumes assemblies to develop from subassemblies through the development of "latent" into "strong" contacts between neurons. In his notation:

$B$ = number of neurons in the brain;

C = maximum number of assemblies in the brain;

N = number of neurons in a subassembly;

y = number of subassemblies in an assembly;

a = number of strong (latent) contacts per neuron.

Legéndy further assumes that half of the strong (latent) contacts make output (axonal) connections and the other half make input (dendritic) connections. Then he defines

m = maximum number of strong contacts from an assembly to one of its subassemblies.

The number of output contacts from an assembly is $(\frac{1}{2} Nya)$ and those connecting to a subassembly reach a fraction $N/B$ of the neurons in the brain. Therefore

$$m = \frac{N^2 ya}{2B} \qquad (7.2.29)$$

The maximum capacity of the brain is reached when about one half of the latent contacts have become strong; thus

$$C = \frac{Ba}{2my} \qquad (7.2.30)$$

Substitution of (7.2.29) into (7.2.30) gives an estimate for the maximum number of assemblies that can be stored in the brain as

$$C = (\frac{B}{Ny})^2 \qquad (7.2.31)$$

This estimate is insensitive to the maximum number of strong contacts assumed for each neuron and to the number of subassemblies. Taking $B = 10^{10}$ neurons in the brain, $N = 10^4$ neurons/subassembly and $y = 30$ subassemblies/assembly gives $C = 10^9$ assemblies.

Considering how much more complex the "multiplex neuron" of Fig. 6-3 is than the simple representation employed by Legéndy, this may be a conservative estimate for "the number of elementary things the brain can know", but, as he points out $10^9$ is the number of seconds in 30 years. Griffith (1971) has reviewed various estimates for the storage capacity of the brain that indicate a lower bound of $10^8$ bits (about equal to the information content of the <u>Encyclopedia Britannica</u> estimated at one bit per character) and an upper bound of $10^{11}$ bits (corresponding to an average rate of 50 bits/sec for 60 years). He feels that a reasonable value lies in the range from $10^9$ to $10^{10}$ bits.

## 3.   SPATIAL EFFECTS

Up to this point we have ignored spatial organization of the neural interconnections. This organization is evident from Fig. 1-4, and such points as: how an auditory reverberation in area 22 could communicate or interact with an optical reverberation in area 19 are now considered here. Beurle (1956) initiated the analysis of wave effects by assuming the "neural medium" to be composed of $\rho$ threshold elements per unit volume and for which:

$\beta(x)dx$ = the average number of connections from one cell to an infinite plane of thickness $dx$ and distance $x$ away;

$F$ = the "activity" or the fraction of cells becoming active per unit of time;

$R$ = the fraction of cells that are sensitive, (i. e. , not refractory);

$\Phi$ = the probability of a cell being stimulated above threshold per unit time;

$\tau$ = the operating (synaptic) delay.

Assuming that the cells after firing remain in a refractory state (unable to fire) for a time period that is longer than the duration of the disturbance being considered, each firing diminishes the fraction of sensitive cells, R. Thus

$$\frac{\partial R}{\partial t} = -F \qquad (7.3.1)$$

The firing rate at $t + \tau$ is equal to R times the probability of a sensitive cell being stimulated above threshold. Thus

$$F(t + \tau) = R(t)\,\Phi(t)$$

or, as in (7. 2. 26)

$$\tau \frac{\partial F}{\partial t} = (R \Phi - F) \tag{7. 3. 2}$$

Partial derivatives are indicated in (7. 3. 1) and (7. 3. 2) because dependence on distance (x) is considered as well as on time (t). Beurle assumed $\beta(x)$ to be an exponentially decreasing probability of interconnection and showed

$$\Phi = mF \tag{7. 3. 3}$$

where m is a real proportionality constant.   More generally

$$\Phi = p(\theta - 1) \int_{-\infty}^{\infty} F(x', t)\beta(x - x')dx' \tag{7. 3. 4}$$

$$= p(\theta - 1)F \otimes \beta \tag{7. 3. 4'}$$

where $\otimes$ indicates the convolution operation in (7. 3. 4) and $p(\theta - 1)$ is the probability of a cell being just one input pulse below threshold. If $p(\theta - 1)$ is assumed constant and the spatial extent of $\beta(x)$ is small compared with that of the disturbance being considered, (7. 3. 4) reduces to (7. 3. 3) with m equal to p times the area under $\beta(x)$.

Here we assume (7. 3. 3) and suppose F and R to be the traveling wave of information with velocity u shown in Fig. 7-10. Then

$$F(x, t) = F(x - ut) = F(\xi)$$

$$R(x, t) = R(x - ut) = R(\xi)$$

where

$$\xi = x - ut \tag{7. 3. 5}$$

is the space coordinate in a frame moving with the assumed velocity u.   Equations (7. 3. 1) and (7. 3. 2) then become

$$u \frac{dR}{d\xi} = F \tag{7. 3. 6a}$$

$$u\tau \frac{dF}{d\xi} = F(1 - mR) \qquad (7.3.6b)$$

or

$$\frac{\tau dF}{dR} = 1 - mR \qquad (7.3.7)$$

which can be integrated over R to obtain the parabolic trajectory

$$\tau F = K + R - \frac{m}{2} R^2 \qquad (7.3.8)$$

sketched in Fig. 7-10d. Given a value, $R_1$, for the fraction of sensitive cells ahead of the wave, the corresponding value in the wake is

$$R_2 = \frac{2}{m} - R_1 \qquad (7.3.9)$$

as indicated in Fig. 7-10d, but it is unstable against growth to $R_1$ or collapse to zero. This instability appears, just as in Fig. 7-7, because no inhibitory interconnections have been assumed, and it is related to the fact that the velocity (u), at which the wave of information propagates, is not determined by the calculation leading to (7.3.8). This velocity is limited by the spatial extent of the interconnection probability, $\beta(x)$, divided by $\tau$; and Smolyaninov (1970) has discussed techniques for calculating the velocities of rectangular pulses of activity through nets with various forms for $\beta(x)$. More recently, Pastushenko (1975) has used an analysis similar to that described in Section 4-4 to find velocities for both a stable wave (higher velocity) and an unstable wave (lower velocity).

Beurle (1956) indicated how a nonlocal (i. e., multiple maxima) character of $\beta(x)$ might stabilize a wave, and went on to consider mechanisms for storing time-sequential memories, such as those described by Penfield and Perot (1963), in a neural mass. As Fig. 7-11a shows, the intersection of two waves will be a moving region of high activity along which, it is assumed, the threshold of excitation will be reduced for those particular neurons which participate in the activity. Eventually (Fig. 7-11b), wave $W_A$ alone will excite a replica of wave $W_B$. In Fig. 7-11c a schematic mechanism is indicated whereby an external wave, $W_{ext}$, in introduced through incoming sensory fibers, $G_{ext}$. This wave propagates across the neural medium and, between planes H and J,

is encoded onto specific internal feedback fibers, R.  These re-
introduce an internal wave, $W_{int}$, which interferes with a subsequent $W_{ext}$  (Fig. 7-11a) to inscribe an appropriate engram of
lowered threshold for particular cells located near the line  L.
Eventually, it is proposed, the wave $W_{int}$  will excite a replica
of the <u>subsequent</u> wave $W'_{ext}$, just as $W_A$  excites a replica of
$W_B$  in Fig. 7-11b.  This $W'_{ext}$ will then stimulate $W'_{int}$, which
in turn stimulates a following $W''_{ext}$, and so forth.  In this manner, the neural mass could "sing out" a temporal sequence of
commonly experienced activity.

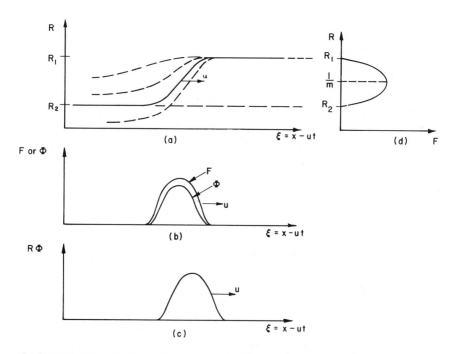

FIGURE 7-10.    Propagation of a simple wave of information in a
                neural medium.

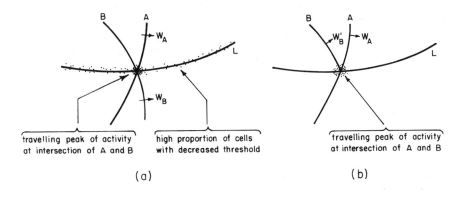

travelling peak of activity
at intersection of A and B

high proportion of cells
with decreased threshold

(a)

travelling peak of activity
at intersection of A and B

(b)

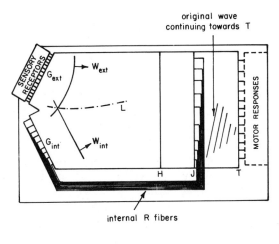

(c)

FIGURE 7-11.    (a) Interaction between two traveling waves of
information; (b) excitation of a replica of $W_B$ by
$W_A$; (c) schematic neural mechanism of a time
sequential memory (see text for details). (From
Beurle, 1956).

Beurle's proposals are somewhat similar to more recent discussions of "holographic" information storage in the cortex (van
Heerden, 1963; Blum, 1967; Longuet-Higgens, 1968; Gabor, 1968a,b,
1969; Pribram, 1969, 1971; Borsellino and Poggio, 1972). These,

in turn, are related to quasiharmonic modal representations of cell assemblies developed by Green (1962) and by Ricciardi and Umezawa (1967) and experimentally demonstrated by Scott (1971). Another approach to the problem of wave propagation in the neural mass was taken by Griffith (1963a, b, 1965, 1971), who pointed out that if only a small fraction of the cells are activated by a wave, its propagation might be described by a linear operator, L(F), supported by a nonlinear (sigmoid) source term, S(F). Thus he investigated the equation

$$L(F) = S(F) \tag{7.3.10}$$

with the linear operator arbitrarily chosen to be a second-order differential operator of the form

$$L = a + b \frac{\partial}{\partial t} + c \frac{\partial^2}{\partial t^2} + d \frac{\partial^2}{\partial x^2} + \frac{\partial^2}{\partial y^2} + \frac{\partial^2}{\partial z^2} \tag{7.3.11}$$

A much more detailed analysis has recently been published by Wilson and Cowan (1973) that assumes subpopulations of excitatory and inhibitory neurons to be described by the firing rates $E(x, t)$ and $I(x, t)$. The result of their study was an augmentation of (7.2.26) to the form

$$
\begin{aligned}
\tau \frac{\partial E}{\partial t} &= -E + (1 - rE)S_e[\rho_e E \otimes \beta_{ee}(x) - \rho_i I \otimes \beta_{ie}(x) + P] \\
\tau \frac{\partial I}{\partial t} &= -I + (1 - rI)S_i[\rho_e E \otimes \beta_{ei}(x) - \rho_i I \otimes \beta_{ii}(x) + Q]
\end{aligned}
\tag{7.3.12a,b}
$$

in which $\beta_{xx}(x)$, $\beta_{ie}(x)$, $\beta_{ei}(x)$, $\beta_{ii}(x)$ are the "excitatory-excitatory," and so forth, interconnection probabilities and $\otimes$ stands for the convolution operation defined in (7.3.4). Equations (7.3.12) evidently reduce to (7.2.26) when the $\beta$ functions have a spatial extent that is small compared with the disturbance under consideration.

As Wilson and Cowan have pointed out, an unsatisfactory feature of (7.3.12) is the inclusion of only one space dimension rather than the two dimensions, which seem appropriate for the description of activity on the cortex surface. Nonetheless, numerical studies of (7.3.12) with physiologically reasonable assumptions for the constants and connection probabilities have

already yielded a variety of interesting results, some of which
are reproduced from Wilson and Cowan (1973) in Fig. 7-12.

In Fig. 7-12a a localized transient is observed in response
to the localized stimulation

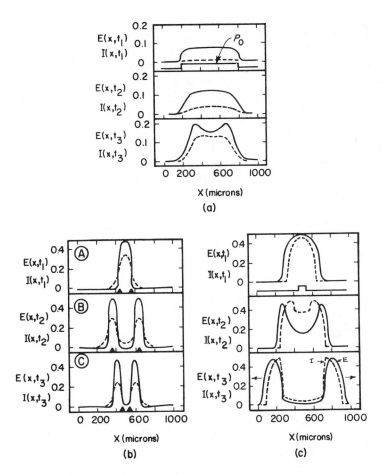

FIGURE 7-12.  Numerical solutions of the Wilson-Cowan equations
(7.3.12): (a) local disturbance; (b) example of binocular fusion; (c) wave of excitation and inhibition (Wilson and Cowan, 1973).

$$P(x, t) = 0 \quad \text{for} \qquad x < 200\,\mu$$
$$x > 800\,\mu$$
$$t > t_1$$

$$= P_o \quad \text{for} \quad 200\,\mu < x < 800\,\mu$$
$$0 < t < t_1$$

This transient exhibits a latent-edge enhancement effect.

In Fig. 7-12b the steady-state response is observed for local-ized inputs at two adjacent points. Once a double response has formed (B) it can be maintained at a much smaller input separa-tion (C) than originally at (A). This "hysteresis" effect closely simulates that observed by Fender and Julesz (1967) of binocular fusion.

In Fig. 7-12c Q has been set to a constant positive level and a localized input is observed to stimulate outward traveling waves. These waves are qualitatively similar to Fitz Hugh-Nagumo waves shown in Fig. 4-10 with $I(x, t)$ playing the role of "recov-ery variable." They differ from Beurle's waves (Fig. 7-10) be-cause the refractory time $(r)$ is short compared with the duration of the disturbance; thus a single neuron can fire many times dur-ing passage.

## 4.  THE ELECTROENCEPHALOGRAM  (EEG)

Elul (1972) has indicated in his recent review that the ele-mentary generators of the EEG are not yet known, although there is substantial evidence that it derives from wave activity of neu-rons in the cortex.  He states:

> Analysis of the relationships between the gross EEG and the wave activity of individual nerve cells indicates that the gross activity is produced through summation of the synchronized activity of a comparatively small fraction of the cerebral neuronal population.  Although the rest of the population also is active, their contributions are not synchronized and therefore summate much less effectively, probably not reaching the limit of resolution of EEG re-corders (i. e. , 1-2 μV).  There is only sketchy information

on the mechanisms of synchronism, but it is clear that
they involve subcortical drives and entail sequential
activation of different groups of cortical neurons.

In view of this uncertainty in the elementary structure, attempts to model the EEG using the linear diffusion equation
(Hendrix, 1965; Anninos and Raman, 1975) or a linearized version
of (7. 3. 10) (Nuñez, 1974a, b) should be considered with caution.
Much more promising, in my opinion, are numerical studies in
which an attempt is made to model physiological data.  Studies
by Anderson, Gillow, and Rudjord (1966) and by Lopes da Silva,
Hocks, Smits, et al. (1974) may help to clarify the role of the
thalamus in generation of the alpha rhythm.
     Due to the intermittent synchronization of cortical neurons,
Elul (1972) emphasizes that it may be inappropriate to treat the
EEG as a stationary statistical process.  "But," he concludes,
"it may not be altogether unrealistic to hope that an array of several hundred cortical electrodes may provide not only a detailed
map of cortical potential distribution in space and time, but also
a meaningful physiological correlate of perception and consciousness. "

# 8

# Knowledge of the Mind

For the green of the leaf that fluttered down on my
hand chlorophyll had to be substituted, and this, like
all that which was said to me of the biochemical find-
ings about the life of the tree, drew me into the world
of $\underline{x}$ where there existed only that which could not be
realized.  Even the space in which the linden was fixed
was unrealizable mathema.  But I put up with it, I accepted the thing or unthing, which had become property-
less and uncanny, the thing that had waited for me in
order to become once again the blooming and fragrant
linden of my sense world.  I said to the sense-deprived
linden-x what Goethe said to the fully sensible rose:
"So it is you. "

Martin Buber

## 1.   TOWARD AN INDUCTIVE STRATEGY

In Chapter 1 was mentioned the hierarchical structure exhibited by modern scientific knowledge.  The fact that this structure is
a present-day phenomenon must be accepted; whether it will necessarily remain as scientific knowledge continues to evolve is
a question for which scientists provide no unified answer.  There
is certainly a tendency to consider some levels of knowledge as
more "fundamental" than others.  Such levels are generally those
in which the reductive and analytical techniques of classical science have proven most successful.  Scientists working at other
levels of the hierarchy often evaluate themselves according to the

262

degree with which they can imitate the more fundamental levels. Psychology and sociology, for example, are often considered "soft" studies because they lack simple basic laws from which all relevant phenomena can be derived.   For those who feel that all knowledge will eventually be related to the logical foliation of a few fundamental laws (characterized, perhaps, by a parsimonious and aesthetic appeal that reflects their divine nature), these soft studies are simply incomplete.   As they mature, many feel, they will increasingly assume the character of "real" science. But there is an ingenuous air surrounding this point of view that is difficult for many scientists to accept (Weyl, 1949; Bohr, 1950, 1958; Bronowski, 1966; Platt, 1966; Elsasser, 1958, 1966, 1969; Blackburn, 1971; Mehra, 1973; Wheeler, 1974; Weizenbaum, 1976).   One obvious reason for the successes of classical physics is that eighteenth- and nineteenth-century physicists directed their attention toward the simplest outstanding problems; they chose the easy paths and claimed these to be fundamental.   It is unfortunate, in my view, that this attitude tends to set the style for scientific efforts at other levels.

General scientific activity can be resolved into inductive and deductive components.   Inductive science is creative.   It requires the imagination of new "paradigms" (perceptions, notions or beliefs) that are more acceptable than former ones (Kuhn, 1962).   In classical science, most of the professional activity is deductive; the logical implications of currently accepted ideas are exhaustively explored.   At levels 6 and 7 of the hierarchical scheme outlined in Chapter 1, however, no widely acceptable paradigms are available; thus the activity must remain primarily inductive.

Suppose that our aim is to induce a theoretical basis for level 6 that is based on classical physics.   There are at least two impediments that we must evaluate and try to overcome; namely indeterminacy and complexity.   Let us take them in order.

a.   Indeterminacy.

A predictive theory of dynamic behavior requires that it be possible (in principle, at least) to make measurements of initial data from which the future course of events can be computed.   Difficulties arise, however, when a prediction is made of the behavior of a cognitive system that, in turn, learns of the prediction. Since the prediction then becomes part of the initial data, it is not

necessarily valid.  For instance, if I decide to prove a behaviorist wrong, any "prediction" he makes of my future activity can
serve as a key element in my choice.  A "prediction" of social
catastrophe may induce compensating countermeasures.  "Self-
fulfilling prophecies" may fool the developmental psychologists.

A second source of indeterminacy at level 6 can arise from
linguistic ambiguity.  An example of the sort of logical contradiction that appears in linguistic systems is given by the statement:
"This statement is false."  If we assume it is false, then it
is obviously true; but if we assume it is true, it is easily proven
to be false.  The contradiction arises because the statement refers
to itself.  Bronowski (1966) has emphasized that any reasonably
rich linguistic system contains self-references and is, therefore,
potentially self-contradictory.  This observation is closely related to a skeleton in the closet of twentieth-century science, namely, Gödel's (1931) incompleteness theorem.  This theorem demon-
strates that arithmetic cannot be given a satisfactory axiomatic
basis.  If such axioms are assumed to permit the proof of all true
theorems, it will be possible to prove false theorems; and if the
axioms are restricted so that no false theorems can be proved, then
there will always be true theorems that cannot be proved.  Although
the implications of Gödel's theorem for the "man-machine" con-
troversy have probably been overemphasized (Nagel and Newman,
1958; Arbib, 1964; George, 1972), the implications for the style of
scientific endeavor have been almost completely ignored.  We
must admit that the various languages employed in human inter-
course are as rich as arithmetic and are riddled with self-reference.
Since the idea that all truth can be developed from a few postu-
lates does not even apply to arithmetic, there is very little justification (beyond a vaguely sentimental desire for certainty in the
midst of chaos) to expect the knowledge of life to be organized in
this way.

Predictive indeterminacy is not unfamiliar to physical scientists.  To compute the trajectory of a classical particle (a baseball, rocket ship, planet, etc.) it is necessary to measure both
its position and its velocity at some instant of time.  But for very
small particles (e.g., electrons) measurement of position destroys
knowledge of velocity and vice versa.  This palpable failure of
classical science has been elevated by physicists to the "Heisen-
berg principle of indeterminacy."  According to Herman Weyl (1949),

Niels Bohr was:

> inclined to widen the domain of uncertainty by adding a
> specific biological principle of indeterminacy (the precise
> content of which is still unknown) to Heisenberg's well-
> established quantum-mechanical principle of indeterminacy.
> He has pointed out in this connection that an observation of
> the state of the brain cells exact enough for a fairly definite
> prediction of the victim's behavior during the next few
> seconds may involve an encroachment of necessarily lethal
> effect -- and thereby make the organism predictable indeed.
> Bohr maintains that in this way analysis of vital phenomena
> by physical concepts has its natural limits; just as one had
> to put up with complementarity as expressed by Heisenberg's
> principle of indeterminacy in order to explain the stability
> of atoms, so are further renouncements demanded of him
> who tries to account for the self-stabilization of living
> organisms.

Skinner makes the point (and I agree) that "neurology" and
"behaviorism" should be considered as distinct levels in the
scientific hierarchy, and the cognitive difficulties of one should
not be permitted to impede the other.   Yet even in terms of the
behaviorist, "complete initial data" implies knowledge of all
stimulations and conditionings throughout the life of the organism.
Although it is certainly possible to predict something with limited
initial data, substantial data difficulties are necessarily involved
in the prediction of everything.   As Bohr (1958) put it:

> The decisive point is that, if we attempt to predict what
> another person will decide to do in a given situation, not
> only must we strive to know his whole background, including
> the story of his life in all respects which may have con-
> tributed to form his character, but we must realize that what
> we are ultimately aiming at is to put ourselves in his place.

In response to such difficulties, Skinner (1969) lists a dozen
indications of technical progress in the experimental analysis of
behavior over the past 35 years.   These include the following:

1.   "Experiments last, not for an hour, but for many hours,
     days, weeks, or even months. "

2.    "The past history of the organism is more carefully controlled, possibly from birth. "

3.    "Many more species have been studied, including man (retardates, psychotics, normal children, and normal adults). "

4.    "Many more reinforcers have been studied -- including, in addition to food and water, sexual stimulation, the opportunity to behave aggressively, and the production of novel stimuli. "

5.    "The experimental space often contains two or more organisms with interlocking contingencies which generate 'synthetic social relations'. "

Such experimental difficulties introduce the second impediment to establishing a classical science at level 6 (Chapter 1), discussed below.

b.    Complexity.

Platt (1966) suggests that if dynamic complexity could be translated into visible brightness "An earthworm would be a beacon, a dog would be a city of light, and human beings would stand out like blazing suns of complexity, flashing bursts of meaning to each other through the dull night of the physical world between. " A scintillating metaphor, the skeptic might comment, but what does it mean? Can complexity be measured? Is there any reason to suppose (in the age of the digital computer) that the behavior of complex dynamic systems cannot be predicted just as well as simple systems?

In attempting to answer such questions, we must remember that at level 6 we are engaged in an inductive activity. To determine regularities of behavior, it is necessary to run a number of experimental trials that is large compared with the number of possible experimental results. Take, as a simple example, the problem of determining whether a particular die is "loaded. " In any given trial, there are six possible results; thus six, twelve, or eighteen rolls will tell almost nothing about the relative tendencies of the faces to appear. But 6000 or 6,000,000 rolls will. Another approach, of course, would be to cut the die into thin slices and analyze each for nonuniformities of density. In this

way we would obtain the answer at the cost of destroying the subject of study.   The difficulty at level 6 is that one very readily
encounters dynamic systems for which the number of distinctly
different modes of behavior becomes too large to handle.

How large is "too large"?   This question has been carefully
studied over the past 20 years by Elsasser (1958, 1966, 1969) in
an "elaboration and substantial extension of Bohr's germinal
ideas. "  He uses the notion of a finite universe to determine a
number  $M \sim 10^{110} \sim 2^{365}$  that is the product of the number of pro-
tons in the universe times the number of picoseconds in the age of
the universe.   Numbers large relative to  M  are called <u>immense</u>,
and any set with an immense number of elements cannot be studied by statistical methods.   A similar notion has independently
been suggested by Bremermann (1967) as a fundamental limit of
computing machines.

Consider now (7. 2. 14), which gives the number of state diagrams that can be formed from  N  arbitrary Boolean elements with
N  inputs.   For N = 1, $\mathfrak{N}_B$ = 4  state diagrams that are easily
sketched.   But suppose we ask how large  N  must be for $\mathfrak{N}_B > M$.
The answer is 6.   Only six model neurons (each of which can
compute an arbitrary Boolean function of the states of the six neurons) can exhibit an immense number of state diagrams!   If the
neuron computations are restricted to threshold discriminations,
as in (7. 2. 15), then nine or ten elements are required to obtain an
immense number of state diagrams.

Suppose that we have before us a set of "black boxes, " each
of which contains a dozen or so interconnected model neurons.
There are at least two reasons why the immense number of possible state diagrams might be of little practical interest.   First, we
might know exactly how the model neurons are interconnected and
what each of them computes.   This would allow us to precisely
determine the particular state diagram embodied by each box.   In
this case the boxes could be considered as "machines" in the
normal sense of the word.   Out of an immense number of possible
configurations, each is nonetheless quite simple and well understood.   Second, the various state diagrams might be functionally
equivalent for the application of interest.   We might, for example,
use the boxes to generate flashing lights at an art show.   Then
only the average firing rates of the neurons would be of interest,
and many of the state diagrams would be functionally equivalent.
Or we could use them as paperweights, making all of the state

diagrams functionally irrelevant.   But, on the other hand, we
might be in the position of not knowing how the neurons in the
boxes are connected and wishing to understand the dynamics of
the corresponding state diagrams.   If we are not allowed to open
the boxes, the only recourse is to investigate and record its dynamic behavior.   Statistics on investigated boxes would tell little
about those that have not been studied.

Perhaps a comment is appropriate here on the definition of a
"machine. "   The word is often used to imply a causal and deter-
minate system of known structure for which the behavior can be
predicted.   But when the term appears in a phrase such as "The
brain is merely a meat machine, " it seems to imply no more than
a dynamical system of undetermined predictability and complexity.
The present author has no quarrel with either usage if the intent
is clear, but it is a long leap to assume that rejecting belief in
magic automatically eliminates mystery.   Nowadays, indeed, dynamical systems constructed by engineers are not necessarily
understandable in times sufficiently short to be of practical interest (Wiener, 1960; Weizenbaum, 1972).

Let us return to the problem of constructing a predictive theory
for the mind of a particular human subject.   In addition to isolating the subject from all knowledge of our predictions, we must
communicate in a language that has been purged of inconsisten-
cies.   The ambiguity of poetry (Empson, 1947) would be tabu, and
even puns might cause difficulty.   Finally, of course, we would
have to investigate the behavior of the subject in some nondestructive way.   Can we expect an immense number of interesting
behavioral sequences?   To answer this, suppose the subject to
have  C  cell assemblies, each representing a particular face,
word, picture, idea, and so on.   Consider the counting of phase
sequences in which  n  assemblies are activated:

$n = 0$    1 state (comatose)

$n = 1$    C  states (each representing the activation of a
            single assembly)

For  $n = 2$  we must form all possible pairs of assemblies.   With
each assembly, $(C-1)$ others can be paired, but the product
$C(C-1)$ counts each pair twice.   Thus:

$n = 2$     $\frac{1}{2} C(C-1)$  sequences (similes, metaphors)

$n = 3$     $\frac{1}{2 \cdot 3} C(C-1)(C-2)$  triplet sequences

$n$     $\frac{C!}{n!\,(C-n)!} = \binom{C}{n}$  n-fold sequences

The sum of <u>all</u> of these sequences

$$\sum_{n=1}^{C} \binom{C}{n} = 2^C$$

is immense for $C > 365$ assemblies; but this is not an important observation because thought trains involving large numbers of perceptual elements are impractical. One must sleep. Much more interesting is the question regarding how many assemblies we must include in a phase sequence for the number of sequences to become immense. For what $n$ is $\binom{C}{n} > M$? This question is readily asnwered since for $n \ll C$

$$\binom{C}{n} \sim \left(\frac{C}{n}\right)^n$$

For $C = 10^9$ [as was estimated from (7.2.31)], $\binom{C}{n}$ is immense when $n > 14$. Thus Roszak's thought train (quoted above) appears to be one of an immense set. A more significant example is presented at the end of this book.

The difficulties besetting those attempting to esatblish psychology as a classical science are analogous to the problems faced by physicists in developing a dynamics for subatomic particles. This analogy is more than superficial, as DeLuca and Termini (1971) have emphasized; there is "a very strong relation between the new epistemological problems raised by cybernetics and quantum mechanics." In both cases the aim is to construct a dynamical theory at a level of the scientific hierarchy that lies obscured by mists of uncertainty. The theory should predict those properties that have been and can be observed. For families of subatomic particles, however, the corresponding mass spectra provide precisely known numerical ratios that are appropriate theoretical goals. Any dynamical theory with stable stationary states characterized by fixed quantities that exhibit the ratios of a known mass spectrum would be of scientific interest. Psychological theorists, on the other hand, do not have a finite set of numbers to predict. Their analog of the mass spectra should be the full range of human behavior. Although this includes the facts of experimental psychology, it is not limited to them.

## 2. A SCIENCE OF THE MIND?

Over the past decade Stephen Grossberg has published an extensive series of papers attempting to provide a mathematical structure for some experimental facts of psychology. Those unfamiliar with this work are advised to consult three carefully written reviews (Grossberg, 1969a, 1974, 1975), from which a brief introduction to his basic ideas is cited here.

Grossberg (1969a) begins by considering the problem of teaching a list of letters of "events" to a "learning subject" M. He emphasizes the observation of "behavioral atoms" as follows:

> If we wish to understand our usage of such simple verbal units as A, we must take seriously our impression that A is a <u>single</u> unit that is never decomposed in actual speech. We do this by assuming that A is represented in M by a <u>single</u> state.

Thus given any n simple behavioral units $r_i$, $i = 1, 2, \ldots, n$, he defines n points $v_i$ in M to represent these units as follows.

$$
\begin{aligned}
&\qquad\qquad\qquad M \\
r_1 &\rightarrow \quad \cdot v_1 \\
r_2 &\rightarrow \quad \cdot v_2 \\
&\quad\vdots \\
r_n &\rightarrow \quad \cdot v_n
\end{aligned}
\tag{8.2.1}
$$

Grossberg's units are taken to be cell populations, single cells, or patches of membrane (see Schmitt, Dev and Smith, 1976), depending on the context. He proves theorems about learning in essentially arbitrary anatomies in which arbitrary sensory data processing, signal generating rules, and decay laws exist. These theorems describe the interaction of short-term memory (STM) traces $x_i(t)$ and long-term memory (LTM) traces $z_{jk}(t)$. The STM traces represent the activity of the behavioral units, and the LTM traces encode associations among the units. Differential equations of the following form describe their dynamic interaction during a learning experience:

$$\dot{x}_i = A_i x_i + \sum_{k=1}^{n} B_{ki} z_{ki} + C_i(t) \qquad (8.2.2)$$

$$\dot{z}_{jk} = D_{jk} z_{jk} + E_{jk} x_k \qquad (8.2.3)$$

where i, j, k = 1, 2, ..., n. The terms $A_i$, $B_{jk}$, $D_{jk}$, and $E_{jk}$ can be chosen quite generally without preventing unbiased pattern learning. These terms have the following heuristic interpretations:

1. The $A_i x_i$ in (8.2.2) represents the decay of an unstimulated behavioral unit; for example, choose $A_i$ a negative constant ($\sim -1 \ \sec^{-1}$ if the $v_i$ are cell assemblies).

2. $C_i(t)$ in (8.2.2) is the experimenter's input to stimulate $x_i$. $C_A(t)$, for example, might represent visual or auditory presentation of the letter A.

3. The term $\sum_{k=1}^{n} B_{ki} z_{ki}$ represents a sum of signals from all units to $v_i$. Each summand $B_{ki} z_{ki}$ represents a signal $B_{ki}$ from $v_k$ to $v_i$ that is gated, or shunted, by the LTM trace $z_{ki}$ before it reaches $v_i$. $B_{ki}$ can be chosen in various ways; for example, it can be a linear function of $x_k(t - \tau_{ki})$ above a signal threshold $\theta_{ki}$, as in $B_{ki}(t) = \alpha_{ki} \sigma [x_k(t - \tau_{ki}) - \theta_{ki}]$. Note that $\alpha_{ki}$ is not necessarily equal to $\alpha_{ik}$.

4. $z_{jk}(t)$ in (8.2.3) measures the slow growth of an interaction between behavioral units $v_j$ and $v_k$ as a result of learning.

5. The $D_{jk} z_{jk}$ term in (8.2.3) represents spontaneous changes in $z_{jk}$ in the absence of learning; for instance, exponential decay of LTM strength if $D_{jk}$ is a negative constant.

6.    The term $E_{jk}x_k$ in (8. 2. 3) implies a growth in interaction strength between the behavioral units $v_j$ and
$v_k$ . The term $E_{jk}$ describes a signal from $v_j$ that
drives this learning process; for example, choosing $E_{jk}$
proportional to $B_{jk}$ suffices in simple cases, but $E_{jk}$
can have a lower threshold, different time lag, and so
on, than $B_{jk}$ without invalidating the theorems about
pattern learning.

Equation (8. 2. 2) might be considered as a representation of
(7. 2. 26) or of the dynamic equations developed by Anninos, Beek,
and Csermely (1970) for the statistical interaction of cell assemblies. With (7. 2. 23), the activity of the ith assembly at time
$t + \tau$ is given by

$$F_i(t + \tau) = [1 - F_i(t)] \, S_i[E_i(t + \tau), \, F_i(t)] \qquad (8.\,2.\,4)$$

where $E_i(t + \tau)$ is the net excitation from other assemblies and
$S_i[\cdot\,,\cdot\,]$ is evaluated by summing from $\theta - E_i$ rather than $\theta$ as
in (7. 2. 16). Anninos and coworkers assume

$$E_i(t + \tau) = \sum_j K_{ij} \, F_j \qquad (8.\,2.\,5)$$

where the $K_{ij}$ are linear coupling coefficients between assemblies. To date they have only discussed pairs of assemblies.
Grossberg's analysis shows how such systems can learn spatial patterns; that is, input vectors of the form $C_i(t) = \alpha_i C(t)$,
where each $\alpha_i \geq 0$ and $\sum_{k=1}^n \alpha_k = 1$. His theorems describe
how the stimulus sampling probabilities

$$Z_{jk} = z_{jk} \left( \sum_{m=1}^n z_{jm} \right)^{-1}$$

converge to the pattern weights $\alpha_k$ as a result of learning. This
is a generalized version of the Perceptron.
Grossberg also discusses aspects of how M can learn a
serial list $\{v_1, v_2, \ldots, v_n\}$. Then $Z_{12} \approx Z_{23} \approx \cdots \approx Z_{n-1,\,n} \approx 1$
and the chain $\{v_1, v_2, \ldots, v_n\}$ has been "embedded into the field
of M's alternatives. " For this reason, Grossberg calls it the
theory of embedding fields. Using this theory [i. e. , (8. 2. 1-3)],

he has been able to display several phenomena of behavioral psychology (Grossberg, 1974, 1975) including: (a) bowing, skewing, chaining, and chunking in the serial learning lists and (b) many aspects of operant conditioning. The heuristic technique used to develop his theory is called "the method of minimal anatomies" and is described by Grossberg (1974) as follows:

> Given specific psychological postulates we derive the minimal network of embedding field type that realizes these postulates. Then we analyze the psychological and neural capabilities of this network. An important part of the analysis is to understand what the network cannot do. This knowledge often suggests what new psychological postulate is needed to derive the next, more complex network. In this way, a hierarchy of networks is derived, corresponding to ever more sophisticated postulates. This hierarchy presumably leads us closer to realistic anatomies, and provides us with a catalog of mechanisms to use in various situations.

Since each network is "embedded" into the next evolutionary stage, another sense of the term appears (Grossberg, private communication). The basic concept embodied in (8. 2. 1) seems consistent with the reverberatory neural theory outlined in Chapter 7; in particular, the $v_i$ of M can be considered to represent cell assemblies. The problem is to understand the extent of possible comprehension of the ultimate dynamics of their activity.

Behavioral psychologists tend to assume a rather precarious stance in relation to this problem. In his recent book Beyond Freedom and Dignity (1971), Skinner boldly asserts (p. 192) "Man is much more than a dog, but like a dog he is within range of scientific analysis. " To this the eminent neuroscientist John C. Eccles (1973, p. 233) responds "Skinner's theory and the technique of operant conditioning were developed from his experiments on pigeons and rats. Let them be the beneficiaries! " Clearly the matter is in some doubt. But it should not be assumed that the behaviorists are evil men who wish to turn us into social insects; they are motivated by concern for the problems faced by human kind. As Skinner (p. 3) puts it:

> *What we need is a technology of behavior* [italics added.] We could solve our problems quickly enough if we could adjust the growth of the world's population as precisely as we adjust the course of a spaceship, or improve agriculture and industry with some of the confidence with which we accelerate high-energy particles, or move toward a peaceful world with something like the steady progress with which physics has approached absolute zero (even though both remain presumably out of reach). But a behavioral technology comparable in power and precision to physical and biological technology is lacking, and those who do not find the very possibility ridiculous are more likely to be frightened by it than reassured. That is how far we are from "understanding human issues" in the sense in which physics and biology understand their fields, and how far we are from preventing the catastrophe toward which the world seems to be inexorably moving.

From the viewpoint of natural philosophy, however, the issue is not what we want or even what we need but what we can have. Angels, too, could solve our problems quickly enough.

To appreciate the complexity of real mental activity, let us take a very simple example. We return to the notion of a phase sequence of serially excited cell assemblies (Hebb, 1949) and consider how an infant might learn to perceive the triangle EFG shown in Fig. 8-1a. The constituent sensations of the vertices E, F, and G are first supposed to be centered on the retina by eye movement and topologically mapped onto area 17 of the cortex (see Fig. 1-4). Corresponding cell assemblies ⓔ, ⓕ, and ⓖ then develop in area 18 through diffuse (nontopological) connections with area 17. The process of examing the triangle involves a phase sequence of the form

ⓔ → ⓕ → ⓔ → ⓖ → ⓔ → ⓕ → ⓔ → ⓕ → ⓖ → (etc.)

Gradually the subassemblies ⓔ, ⓕ and ⓖ will fuse into a common assembly for Gestalt perception of the triangle ⓣ

ⓔ + ⓕ + ⓖ ≡ ⓣ

with, perhaps, intermediate perceptions of double vertices such as ⓔ + ⓕ. Thus the phase sequence experienced on examining the triangle eventually becomes (say)

$$ⓖ \rightarrow ⓕ \rightarrow ⓕ + ⓖ \rightarrow ⓔ + ⓕ \rightarrow ⓣ \rightarrow ⓔ \rightarrow ⓔ$$

$$+ \quad ⓖ \rightarrow ⓣ \rightarrow ⓕ \rightarrow ⓣ \rightarrow ⓖ \rightarrow ⓣ \rightarrow (etc.)$$

With further development of the assembly ⓣ (which reduces its threshold through strengthening of the internal connections), a glance at a corner plus a couple of peripheral cues can ignite ⓣ; as, for example

$$ⓔ \rightarrow ⓔ + ⓕ \rightarrow ⓔ + ⓖ \rightarrow ⓣ$$

The perception ⓣ is then ready to serve as one of the points $v_i$ in Grossberg's theory.

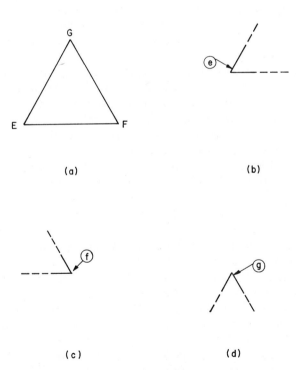

(a)      (b)

(c)      (d)

FIGURE 8-1.   Diagrams related to Hebb's discussion of learning to perceive a triangle.

Those who consider this discussion an unnecessary complication of a simple perceptive event should read von Senden's
Space and Sight (1960).  Originally published in 1932, this unique
account of sixty-five cases, in which congenital blindness from
cataracts had been corrected between the ages of 3 and 46 years,
provided much of the experimental basis for Hebb's theory.  One
of the few uniformities in these cases, writes von Senden, is that
the process of learning to see "is an enterprise fraught with in-
numerable difficulties, and that the common idea that the patient
must necessarily be delighted with the gifts of light and colour
bequeathed to him by the operation, is wholly remote from the
facts. "  Subjects who were entirely dependent on tactile impressions before the operation had an awareness of space (if it can be
so called) that was totally different from a normal visual awareness.  At first, "The patient feels visual impressions to be something alien, intruding on his mind without action on his own part,"
and later, "Given that attention is present, the stimuli impinging
on the visual organ from an objective shape merely occasion the
act of perception as such, but do not determine its outcome . . . .
The final development up to the fully formed idea of shape involves
a series of transitional forms as intermediate stages, which develop one from another, and are liable to vary between individuals,
since it is the individual who himself creates them. "

As constituted in  (8. 2. 1-3), Grossberg's theory might be applied to describe the growth of an assembly from its subassem-
blies.  The development of $\textcircled{t}$ from $\textcircled{e}$ , $\textcircled{f}$ and $\textcircled{g}$ , for example, could be interpreted as a learning of the list  E, F, and  G
in any order.  But a fundamental difficulty is this: after the assembly $\textcircled{t}$ has been formed, a corresponding point [ see (8. 2. 1)]
should be added to the description of  M.  In this sense, Grossberg's early theory lacks the structural dynamics proposed by
Hebb; how it could be embedded into an augmented theory that
would eliminate this problem is not entirely clear.  In the spirit of
the method of minimal anatomies, (8. 2. 1- 3) can also be criticized
because they include only interactions between pairs of assemblies.  An early paper (Grossberg,1969b) discusses the possibility
of using higher-order LTM functions that would compute products
of the activities in three or more assemblies, rather than just in
pairs of assemblies, as in (8. 2. 3).  Such a mechanism creates
enormous numbers of pathways with no addition in the number of
behaviorally coded units, and severe noise problems result.  Later

work abandons this approach and focuses on mechanisms of hierarchical coding, reinforcement, STM, and attention. In Grossberg
(1970, 1972a) are described anatomies in which certain higher-
order cell assemblies will fire only in response to prescribed, but
otherwise arbitrary, patterns of activity on lower-order cell assemblies. These higher-order cell assemblies code the lower-
order patterns. In Grossberg (1971, 1972b) are described mechanisms of reinforcement whereby the activity of cell assemblies
can trigger mechanisms of reinforcement and motivation that interact with drive levels to regulate overt performance, and recently
(Grossberg, 1975) these mechanisms are developed to yield a
theory of attention in which activity in the coded hierarchy of cell
assemblies is continually reorganized based on the system's suc-
cess in achieving prescribed expectations. Similar principles
have shown how to retune, or recode, an assembly's responsive-
ness based on early developmental experiences (Grossberg, 1976a),
and how a behavioral plan, or goal-oriented sequence of sensory-
motor coordinations, can spontaneously be generated in the cell
assembly hierarchy (Grossberg, 1976b, c). Deciding how well
such formal network mechanisms describe some of the complex
processing in vivo will not be easy.

    Blum (1967) describes a class of cortex theories in which the
cell assemblies can communicate by sending and receiving wave
pulses. Spatial regions of constructive interference are then favored for action of active assemblies on inactive ones, and spatial constraints cease to be important in determining the probability of an interaction. The structure of the major assembly, on
which consciousness is directed at a particular moment, is not
limited by any predetermined hierarchical structure. One of Blum's
objectives is to suggest "a new area of analogies, mechanistic in
its primitives, which may be more appropriate to the problem of
understanding central nervous system function" and "open our
eyes to the insidious preconceptions for biology of our particular
mathematical culture, which is based upon centuries of successful involvement with physics." Some consideration has also been
given by Good (1965) to the interaction of assemblies through their
common subassemblies. He suggests that interactions at least as
high as the sixth order should be important, but he finds fundamental difficulties in defining statistical interaction parameters.
However, since his main objective is simply the design of an
"ultraintelligent machine," understanding that machine is of

secondary interest.

In considering the spectrum of possibilities for cell assembly interactions we should not forget the complex capability for information processing that is available to a single neuron. On the axonal side, Chung, Raymond, and Lettvin (1970) have suggested that temporal pulse trains may be mapped onto a subset of the distal branches as a meaningful spatial pattern. And on the dendritic side, arbitrary Boolean functions of the spatiotemporal input patterns may be computed. If individual neurons "speak with meaning" and also participate in cell assembly activity, the assembly interaction could be very sophisticated. A vast arena emerges for "subconscious" mental activity.

Hebb's theory, it should be emphasized, is descriptive rather than computational. He has proposed a reasonable development of the concepts of psychology from the elements of thought; namely, a sketch for level 6. He makes a careful distinction between the concepts of "sensation" and "perception." Sensation is an automatic neural response to sensory input such as, for example, the excitation that appears in area 17 after the vertex E is centered on the retina, whereas perception involves the activation of an appropriate cell assembly. Sensation is determined while perception (see Fig. 7-3) is often ambiguous. This concept of perception should be compared with that of the Perceptron, discussed in Section 7-1, which can rather easily recognize simple geometrical forms (i. e. , circle, square, triangle, letters of the alphabet) but has great difficulty with the same figures appearing in the context of other lines and angles. *

In 1949 Hebb suggested that a phylogenetically significant parameter for evaluating the nature of animal brains is the ratio of association area to sensory area or A/S ratio of the cortex. For lower creatures, such as insects (and currently available computers), this ratio is essentially, zero, but with higher species the A/S ratio grows, bringing both a useful ability to modify motivation according to perceptions and a potentially dangerous tendency to confuse those perceptions with the external world. Thus it seems naive to extrapolate to human behavior from that of a dog.

---

* Perhaps the Perceptron should better have been named the "Sensatron."

## 3. CONCLUDING REMARKS

As at other levels of the scientific hierarchy, psychological facts might be viewed as spatiotemporal entities that form in a sea of cell assemblies. The appropriate "chemistry," however, is vastly more complex than that leading from the atomic elements to the wonders of cytology. In each person's head are a large number of major assemblies; perhaps a billion (or even more) could exist in the mind of a highly developed individual. The nature of these will vary from person to person, from time to time, and from culture to culture; the ways in which they might interact are even more obscure. Our minds are exotic gardens and each may grow different from the next. Indeed, one of the important functions of human culture appears to be the imposition of some order, through training and tabu , on the potential chaos of human experience. By means of this order (Bohr, 1958) "latent potentialities of human life can unfold themselves in a way which reveals to us new aspects of its unlimited richness and variety." In this sense we have had many "technologies of behavior" since the stone age villages. Clearly, mankind now faces a challenge to develop a world culture with appropriate local variations that will permit us to survive and prosper. Although we scientists (particularly the behavioral psychologists) may have something to contribute toward this development, we have not yet shown an appreciation for the complexity of human nature that would qualify us to take charge.

A frightening example of such technological insensitivity is the proposal to build a computing machine that exceeds human intellectual power. In this way the Gordian knot of human mystery would be cut by cool-eyed engineers. A carefully prepared exposition of this proposal is given by Good (1965), who opens with the statement: "The survival of man depends upon the early construction of an ultraintelligent machine. ... " and concludes with the opinion:

It is more probable than not that, within the twentieth century, an ultraintelligent machine will be built and that it will be the last invention that man need make, since it will lead to an "intelligence explosion." This will transform society in an unimaginable way. The first ultraintelligent machine will need to be ultraparallel, and is likely to be achieved with the help of a very large artificial neural net.  The required high degree of connectivity might be attained with the help of microminature radio transmitters and receivers.

Would such a machine really help us understand human experience?  Or would it become emotionally unstable?  Or attempt to reduce us to slaves?  Those who propose such a machine should read again Mary Shelley's Frankenstein (or a Modern Prometheus), where at the end her creature laments:

> I have destroyed my creator, the select specimen of all that is worthy of love and admiration among men,...  There he lies, white and cold in death.

What we face in the investigation of human nature are dynamic processes that (although we suppose them to be entirely natural) are so uncertain, so complex and, at the same time, so interesting that they must be individually experienced.  Maslow (1968) has courageously directed his colleagues toward the depths of these mysteries with his emphasis on the "positive psychology" of fully functioning and healthy human beings.  In his Psychology of Science he presents a strong appeal for scientists to reestablish the primacy of experience as the "basic coin in the realm of knowing."  The aim of human knowledge should not be prediction and control but understanding and self-realization.  As Rogers (1961) put it, the emphasis should be On Becoming a Person.  Just as the elementary particle physicists is forced to renounce the possibility of predicting and controlling trajectories, the behavioral psychologists must admit that important areas of human activity lie beyond his ken.  Behaviorism is not wrong but simply incomplete and irrelevant for understanding such psychic phenomena as the peak experiences of "being cognition" studied by Maslow.  Although the reality of such experiences is attested by every honest poem, there is perhaps no better description than that given by Martin Buber (1970) in recalling his perception of the linden tree:

I can accept it as a picture: a rigid pillar in a flood
of light, or splashes of green traversed by the gentleness
of the blue silver ground.

I can feel it as movement: the flowing veins around
the sturdy, striving core, the sucking of the roots, the
breathing of the leaves, the infinite commerce with earth
and air -- and the growing itself in its darkness.

I can assign it to a species and observe it as an instance, with an eye to its construction and its way of life.

I can overcome its uniqueness and form so rigorously
that I recognize it only as an expression of the law --
those laws according to which a constant opposition of
forces is continually adjusted, or those laws according
to which the elements mix and separate.

I can dissolve it into a number, into a pure relation
between numbers, and eternalize it.

Throughout all of this the tree remains my object and
has its place and its time span, its kind and condition.

But it can also happen, if will and grace are joined,
that as I contemplate the tree I am drawn into a relation,
and the tree ceases to be an It.   The power of exclusive-
ness has seized me.

This does not require me to forego any of the modes
of contemplation.   There is nothing that I must not see in
order to see, and there is no knowledge that I must forget.
Rather is everything, picture and movement, species and
instance, law and number included and inseparably fused.

Whatever belongs to the tree is included; its form and
its mechanics, its colors and its chemistry, its conversation with the elements and its conversation with the stars --
all this in its entirety.

# References

Adam, N. K. , 1921, 1922, "The properties and molecular structure of thin films," Proc. Roy. Soc. (Lond.) Ser. A, 99, 336-351; 101, 452-472 and 516-531.

Adrian, 1914, "The all-or-none principle in nerve," J. Physiol. 47, 460-474.

Agin, D. , 1969, "An approach to the physical basis of negative conductance in the squid axon," Biophys. J. 9, 209-221.

Aiello, A. , E. Burattini, and E. R. Caianiello, 1970, "Synthesis of reverberating neural networks," Kybernetik 7, 191-195.

Albus, J. S. , 1971, "A theory of cerebellar function," Math. Biosci. 10, 25-61.

Amari, S. , 1971, "Characteristics of randomly connected threshold-element networks and network systems," Proc. IEEE 59, 35-47.

Amari, S. , 1974, "A mathematical theory of nerve nets," Adv. Biophys. 6, 75-120.

Andersen, P. , M. Gillow, and T. Rudjord, 1966, "Rhythmic activity in a simulated neuronal network," J. Physiol. 185, 418-428.

Andersson, P. , B. Holmquist, and P. E. Voorhoeve, 1966, "Entorhinal activation of dentate granule cells," Acta Physiol. Scand. 66, 448-460.

284    References

Anderson, P. W., 1972, "More is different. Broken symmetry and the nature of the hierarchical structure of science," <u>Science</u> <u>177</u>, 393-396.

Andronov, A. A., A. A. Vitt, and S. E. Khaikin, 1966, <u>Theory of Oscillators</u>, Addison Wesley, New York.

Anninos, P. A., 1972a, "Mathematical model of memory trace and forgetfulness," <u>Kybernetik</u> <u>10</u>, 165-167.

Anninos, P. A., 1972b, "Cyclic modes in artificial neural nets," <u>Kybernetik</u> <u>11</u>, 5-14.

Anninos, P. A., 1973, "Evoked potentials in artificial neural nets," <u>Kybernetik</u> <u>13</u>, 24-29.

Anninos, P. A., B. Beek, T. J. Csermely, E. M. Harth, and G. Pertile, 1970, "Dynamics of neural structures," <u>J. Theoret. Biol.</u> <u>26</u>, 121-148.

Anninos, P. A. and R. Elul, 1974, "Effect of structure on function in model nerve nets," <u>Biophys. J.</u> <u>14</u>, 8-19.

Anninos, P. A. and S. Raman, 1975, "Derivation of a mathematical equation for the EEG and the general solution within the brain and in space," <u>Int. J. Theoret. Phys.</u> <u>12</u>, 1-9.

Arbib, M. A., 1964, <u>Brains, Machines, and Mathematics</u>, McGraw-Hill, New York, Chapter 5.

Arima, R., and Y. Hasegawa, 1963, "On global solutions for mixed problem of a semilinear differential equation," <u>Proc. Jap. Acad.</u> <u>39</u>, 721-725.

Arnheim, R., 1954, <u>Art and Visual Perception</u>, University of California Press, Berkeley.

Arnold, J. M., W. C. Summers, D. L. Gilbert, R. S. Manalis, N. W. Daw, and R. J. Lasek, 1974, <u>A Guide to Laboratory Use of the Squid LOLIGO PEALEI</u>, Marine Biological Laboratory, Woods Hole, Massachusetts.

Aronson, D. B. and H. F. Weinberger, 1975, "Nonlinear diffusion in population genetics, combustion, and nerve propagation, " Proc. Tulane Program in PDE. Springer-Verlag, New York.

Arshavskii, Yu. I. , M. B. Berkinblit, S. A. Kovalev, V. V. Smolyaninov, and L. M. Chailakhyan, 1965, "The role of dendrites in the functioning of nerve cells, " Dokl. Akad. Nauk SSSR 163, 994-997; translated in Doklady Biophysics, New York.

Arvanitaki, A. , 1940, "Réactions déclenchées sur un axone au repose par l'activité d'un autre axone au niveau d'une zone de contact, " C. R. Soc. Biol. 133, 39-42, 208-215.

Arvanitaki, A. , 1942, "Effects evoked in an axon by the activity of a contiguous one, " J. Neurophysiol. 5, 89-108.

Asanuma, H. , S. D. Stoney, Jr. , and C. Abzug, 1968, "Relation between afferent input and motor outflow in cat motor sensory cortex, " J. Neurophysiol. 31, 670-694.

Ashby, W. R. , H. von Foerster, and C. C. Walker, 1962, "Instability of pulse activity in a net with threshold, " Nature, 196, 561-562.

Averbach, M. S. and D. N. Nasonov, 1950, "The law of self regulation of spreading stimulation ("all or nothing"), " Fiziologicheskii Zhurnal SSR, 36, 46-63 (in Russian).

Balakhovskii, I. S. , 1968, "Constant rate of spread of excitation in an ideal excitable tissue, " Biophysics 13, 864-868.

Barrett, J. N. and W. E. Crill, 1974a, "Specific membrane properties of cat motoneurons, " J. Physiol. 239, 301-324.

Barrett, J. N. and W. E. Crill, 1974b, "Influence of dendritic location and membrane properties on the effectiveness of synapses on cat motoneurons, " J. Physiol. 239, 325-345.

Barron, D. H. and B. H. C. Matthews, 1935, "Intermittent conduction in the spinal cord, " J. Physiol. , 85, 73-103.

286     References

Bennett, M. V. L. , G. D. Pappas, E. Aljure, and Y. Nakajima,
   1967, "Physiology and ultrastructure of electronic junctions,"
   J. Neurophysiol. 30, 180-208.

Berestovskii, G. N. , 1963, "Study of single electric model of
   neuristor, " Rad. Eng. Elec. Phys. 18, 1744-1751.

Berestovskii, G. N. , Ye. A. Liberman, V. Z. Lunevskii, and
   G. M. Frank, 1970, "Optical investigations of change in the
   structure of the neural membrane on passage of a nerve impulse," Biophysics, 15, 60-67.

Berkinblit, M. B. , I. Dudzyavichus, and L. M. Chailakhyan,
   1971, "Dependance of the rate of spread of an impulse in a
   nerve fiber on the capacitance of its membrane," Biophysics
   16, 594-595.

Berkinblit, M. B. , N. D. Vvedenskaya, L. S. Gnedenko, S. A.
   Kovalev, A. V. Kholopov, S. V. Formin, and L. M. Chailakhyan, 1970, "Computer investigation of the features of conduction of a nerve impulse along fibers with different degrees
   of widening, " Biophysics 15, 1121-1130.

Berkinblit, M. B. , N. D. Vvedenskaya, L. S. Gnedenko, S. A.
   Kovalev, A. V. Kholopov, S. V. Formin, and L. M. Chailakhyan, 1971, "Interaction of the nerve impulses in a node of
   branching (investigation of the Hodgkin-Huxley model),"
   Biophysics 16, 105-113.

Bernstein, J. , 1868, "Ueber den zeitlichen Verlauf der negativen
   Schwankung des Nervenstroms, " Arch. ges. Physiol. , 1,
   173-207.

Bernstein, J. 1902, "Untersuchungen zur Thermodynamik der bio-
   elektrischen Ströme, " Arch ges. Physiol. 92, 521-562.

Beurle, R. L. , 1956, Properties of a mass of cells capable of re-
   generating pulses, " Phil. Trans. Roy. Soc. 240 A, 55-94.

Bishop, G. H. 1958, "The dendrite: receptive pole of the neur-
   rone, " Electroenceph. Clin. Neurophysiol. Suppl. 10, 12.

Blackburn, T. R., 1971, "Sensuous-intellectual complementarity in science," Science 172, 1003-1007.

Blair, E. H. and J. Erlanger, 1940, "Interaction of medullated fibers of a nerve tested with electric shocks," Am. J. Physiol. 131, 483-493.

Block, H. D., 1962, "The Perceptron: A model for brain functioning," Rev. Mod. Phys. 34, 123-135.

Block, H. D., B. W. Knight, Jr., and F. Rosenblatt, 1962, "Analysis of a four-layer series-coupled Perceptron," Rev. Mod. Phys. 34, 135-142.

Blum, H., 1967, "A new model of global brain function," Persp. Biol. Med. 10, 381-408.

Bodian, D., 1952, "Introductory survey of neurons," Cold Spring Harbor Symp. Quant. Biol. 17, 1-13.

Bogdanov, K. Yu. and V. B. Golovchinskii, 1970, "Extracellularly recorded action potential and the possibility of spread of excitation over the dendrites," Biophysics 15, 672-681.

Bogoslovskaya, L. S., I. A. Lyubinskii, N. V. Pozin, Ye. V. Putsillo, L. A. Shmelev, and T. M. Shura-Bura, 1973, "Spread of excitation along a fiber with local inhomogeneities (results of modelling)," Biophysics 18, 944-948.

Bohr, N., 1950, "On the notions of causality and complementarity," Science 111, 51-54.

Bohr, N., 1958, Atomic Physics and Human Knowledge, Wiley, New York.

Boltaks, B. I., V. Ya. Vodyanoi, and N. A. Fodorovich, 1971, "Mechanisms of conductivity of phospholipid membranes," Biophysics 16, 856-864.

Bonhoeffer, K. F., 1948, "Activation of passive iron as a model for the excitation of nerve," J. Gen. Physiol. 32, 69-91.

Borsellino, A. and T. Poggio, 1972, "Holographic aspects of temporal memory and optomotor responses, " Kybernetik 10, 58-60.

Boussinesq, M. J. , 1872, "Théorie des ondes et des remous qui se propagent le long d'un canal rectangulaire horizontal, en communiquant au liquide contenu dans ce canal des vitesses sensiblement pareilles de la surface au fond, " J. Math. Pures Appl. 17, 55-108.

Brady, S. W. , 1970, "Boundedness theorems and other mathematical studies of a Hodgkin-Huxley type system of differential equations, " Math. Biosci. 6, 209-282.

Branton, D. and R. B. Park, 1968, Papers on Biological Membrane Structure, Little Brown, Boston.

Brazier, M. A. B. , 1959, "The historical development of neurophysiology, " in Am. Physiol. Soc. , Sec. 1, Vol. I, Handbook of Physiology, 1-58.

Bremermann, H. J. , 1967, "Quantum noise and information, " Proceedings Fifth Berkeley Symposium on Mathematical Statistics and Probability IV, University of California Press.

Bretscher, M. S. , 1973, "Membrane structure: some general principles, " Science 176, 622-629.

Bronowski, J. , 1966, "The logic of the mind," Am. Sci. 54, 1-14.

Buber, M. , 1970, I and Thou, Scribner, New York.

Bullock, T. H. , 1953, Properties of some natural and quasi-artificial synapses in polychaetes, " J. Compar. Neurol. 98, 37-68.

Bungenberg de Jong, H. G. and J. Bonner, 1935, "Phosphatide autocomplex coacervates as ionic systems and their relation to the protoplasmic membrane, " Protoplasma 15, 198-218.

Buratti, R. J. , and A. G. Lindgren, 1968, "Neuristor waveforms and stability by the linear approximation," Proc. IEEE 56, 1392-1393.

Burattini, E. and V. Liesis, 1972, "A method of analysis of the models of neural nets, " Kybernetik 10, 38-44.

Burke, R. E. , 1967, "Composite nature of the monosynaptic excitatory postsynaptic potential, " J. Neurophysiol. 30, 1114-1137.

Butz, E. G. and J. D. Cowan, 1974, "Transient potentials in dendritic systems of arbitrary geometry, " Biophys, J. 14, 661-689.

Caianiello, E. R. , 1961, "Outline of a theory of thought-processes and thinking machines, " J. Theoret. Biol. 1, 204-235.

Caianiello, E. R. , A. de Luca, and L. M. Ricciardi, 1967, "Reverberations and control of neural networks, " Kybernetik 4, 10-18.

Canosa, J. , 1973, "On a nonlinear diffusion equation describing population growth, " IBM J. Res. Dev. 17, 307-313.

Carpenter, G. A., 1974, "Traveling wave solutions of nerve impulse equations, " Ph. D. thesis, University of Wisconsin.

Carpenter, G. A. , 1977a, A geometric approach to singular perturbation problems with applications to nerve impulse equations, " J. Differential Eq. 23, 335-367.

Carpenter, G. A. , 1977b, "Periodic solutions of nerve impulse equations, " J. Math. Anal. Appl., 58, 152-173.

Casten, R. G. , H. Cohen, and P. A. Lagerstrom, 1975, "Perturbation analysis of an approximation to Hodgkin-Huxley theory, " Quart. Appl. Math. , 32, 365-402.

Cereijido, M. and C. A. Rotunno, 1970, Introduction to the Study of Biological Membranes, Gordon and Breach, New York.

Chandler, W. D. , R. FitzHugh, and K. S. Cole, 1962, "Theoretical stability properties of a space clamped axon, " Biophys. J. 2, 105-127.

Changeux, J. P. , 1969, "Remarks on the symmetry and cooperative properties of biological membranes, " A. Engstrom and

B. Strandberg (eds.), <u>Nobel Symposium</u> <u>11</u>, Wiley-Interscience, New York, p. 235.

Changeux, J. P., J. Thiery, Y. Tung, and C. Kittel, 1976, "On the cooperativity of biological membranes," <u>Proc. N. A. S.</u> <u>57</u>, 335-341.

Chizmadzhev, Yu. A., V. S. Markin, and A. L. Muler, 1973, "Conformational model of excitable cell membranes - II. Basic equations," <u>Biophysics</u> <u>18</u>, 70-76.

Chizmadzhev, Yu. A., A. L. Muler, and V. S. Markin, 1972, "Conformational model of excitable cell membranes - I. Ionic permeability," <u>Biophysics</u> <u>17</u>, 1061-1065.

Chung, S. H., S. A. Raymond, and J. Y. Lettvin, 1970, "Multiple meaning in single visual units," <u>Brain Behav. Evol.</u> <u>3</u>, 72-101.

Clark, J. and R. Plonsey, 1966, "A mathematical evaluation of the core conductor model," <u>Biophys. J.</u> <u>6</u>, 95-112.

Clark, J. and R. Plonsey, 1968, "The extracellular potential field of the single active nerve fiber in a volume conductor," <u>Biophys. J.</u> <u>8</u>, 842-864.

Clark, J. W. and R. Plonsey, 1970, "A mathematical study of nerve fiber interaction," <u>Biophys. J.</u> <u>10</u>, 937-957.

Clark, J. W. and R. Plonsey, 1971, "Fiber interaction in a nerve trunk," <u>Biophys. J.</u> <u>11</u>, 281-294.

Cohen, H., 1971, "Nonlinear diffusion problems," in A. H. Taub (ed.), <u>Studies in Applied Mathematics</u>, Prentice-Hall, Englewood Cliffs, New Jersey, 27-64.

Cohen, L. B., B. Hille, and R. D. Keynes, 1970, "Changes in axon birefringence during the action potential," <u>J. Physiol.</u> <u>211</u>, 495-515.

Cohen, L. B. and D. Landowne, 1974, "The temperature dependence of the movement of sodium ions associated with nerve impulses," <u>J. Physiol.</u> <u>236</u>, 95-111.

Cole, K. S. , 1949, "Dynamic electrical characteristics of the squid axon membrane, " Arch. Sci. Physiol. 3, 253-258.

Cole, K. S. , 1968, Membranes, Ions and Impulses, University of California Press, Berkeley and Los Angeles.

Cole, K, S. , 1975, "Neuromembranes: paths of ions, " in G. Adelman (ed. ), Neurosciences: Paths of Discovery, Rockefeller University Press, New York.

Cole, K. S. , H. A. Antosiewicz, and P. Rabinowitz, 1955, "Automatic computation of nerve excitation, " SIAM J. 3, 153-172.

Cole, K. S. and R. F. Baker, 1941, "Longitudinal impedance of the squid giant axon, " J. Gen. Physiol. 24, 771-788.

Cole, K. S. and H. J. Curtis, 1938, "Electrical impedance of nerve during activity, " Nature 142, 209.

Cole, K. S. and H. J. Curtis, 1939, "Electric impedance of the squid giant axon during activity, " J. Gen. Physiol. 22, 649-670.

Cole, K, S. , R. Guttman, and F. Bezanilla, 1970, "Nerve membrane without threshold, " Proc. Natl. Acad. Sci. 65, 884-891.

Cole, K. S. , and J. W. Moore, 1960, "Potassium ion current in the squid giant axon: dynamic characteristic, " Biophys. J. 1, 1-14.

Compton, K. T. and I. Langmuir, 1931, "Electrical discharges in gases, " Rev. Mod. Phys. 3, 191-257.

Conley, C. and R. Easton, 1971, "Isolated invariant sets and isolating blocks, " Trans. Am. Math. Soc. 158, 35-61.

Cooley, J. W. and F. A. Dodge, Jr. , 1966, "Digital computer solutions for excitation and propagation of the nerve impulse," Biophys. J. 6, 583-599.

Cowan, J. D. , 1970, "A statistical mechanics of nervous activity," in Am. Math. Soc. , Some Mathematical Questions in Biology, Providence, R. I. , pp. 1-57.

Cragg, B. G. , and L. H. Hamlyn, 1955, "Action potential of the pyramidal neurons in the hippocampus of the rabbit, " J. Physiol. (Lond. ) 129, 608-627.

Crane, H. D. , 1962, "Neuristor-- a novel device and system concept, " Proc. IRE 50, 2048-2060.

Crane, H. D. , 1964, "Possibilities for signal processing in axon system, " in R. F. Reiss (ed. ), Neural Theory and Modeling, Standord University Press.

Cyrulnik, R. A. , P. A. Anninos, and R. Marsh, 1974, "Complex symptomatology simulated by unstructured neural nets, " Can. J. Neurol. Sci. 1, 17-22.

Danielli, J. F. , 1936, "Some properties of lipoid films in relation to the structure of the plasma membrane, " J. Cell. Compar. Physiol. 7, 393-408.

Danielli, J. F. and H. Davson, 1935, "A contribution to the theory of permeability of thin films, " J. Cell. Compar. Physiol. 5, 495-508.

Dean, R. B. , 1939, "Potentials at oil water interfaces, " Nature 144, 32-33.

Dean, R. B. , H. J. Curtis, and K. S. Cole, 1940, "Impedance of biomolecular films, " Science 91, 50-51.

Deck, K. A. and W. Trautwein, 1964, "Ionic currents in cardiac excitation, " Arch. ges. Physiol. 280, 63-80.

DeLuca, A. , and S. Termini, 1971, "Algorithmic aspects in complex systems analysis, " Scientia 106, 659-671.

Devaux, H. , 1936, "Determination de l'épaisseur de la membrane d'albumine formée entre l'eau et la benzine et proprietes de cette membrane, " C. R. Acad. Sci. 202, 1957-1960.

Donati, F. and H. Kunov, 1976, "A model for studying velocity variations in unmyelinated axons, " Trans. IEEE Biomed. Eng. BME-23, 23-28.

Duncan, D. D. , 1974, Goodbye Picasso, Grosset Dunlap, New York, p. 142.

Dunin-Barkovskii, V. L. , 1970, "Fluctuations in the level of activity in simple closed neurone chains, " Biophysics 15, 396-401.

Eccles, J. C. , 1960, "The properties of the dendrites, " in D. B. Tower and J. P. Schade, eds. ), Structure and Function of the Cerebral Cortex, America Elsevier, New York, pp. 192-203.

Eccles, J. C. , 1964, The Physiology of Synapses, Springer-Verlag, Berlin.

Eccles, J. C. , 1973, The Understanding of the Brain, McGraw-Hill, New York.

Eckhaus, W. , 1965, Studies in Non-Linear Stability Theory, Springer-Verlag, New York.

Ehrenstein, G. , 1976, "Ion channels in nerve membranes, " Physics Today (October), 33-39.

Einstein, A. , 1905, Über die von der molekularkinetischen Theorie der Wärme geforderte Bewegung von in ruhenden Flüssigkeiten suspendierten Teilchen, " Ann. Physik 17, 549-560.

Eisenberg, M. , J. E. Hall, and C. A. Mead, 1973, "The nature of the voltage-dependent conductance induced by alamethicin in black lipid membranes, " J. Membrane Biol. 14, 143-176.

Eleonskii, V. M. , 1968, "Stability of simple stationary waves related to the nonlinear diffusion equation, " S. P. JETP 26, 382-384.

Ellias, S. A. , and S. Grossberg, 1975, "Pattern formation, contrast control, and oscillations in the short term memory of shunting on-center off-surround networks, " Biol. Cybernet. 20, 69-98.

Elsasser, W. M. , 1958, The Physical Foundation of Biology, Pergamon, New York.

Elsasser, W. M. , 1966, Atom and Organism, Princeton University Press.

Elsasser, W. M. , 1969, "Acausal phenomena in physics and biology: a case for reconstruction, " Am. Sci. 57, 502-516.

Elul, R. , 1972, "The genesis of the EEG, " Int. Rev. Neurobiol. 15, 227-272.

Empson, W. , 1947, Seven Types of Ambiguity, New Directions, New York.

Evans, J. , and N. Shenk, 1970, "Solutions to axon equations, " Biophys. J. 10, 1090-1101.

Evans, J. W. , 1972, "Nerve axon equations: I. Linear approximations, II. Stability at rest, III. Stability of the nerve impulse, " Ind. U. Math. J. 21, 877-885; 22, 75-90, and 755-593.

Evans, J. W. , 1975, "Nerve axon equations: IV. The stable and the unstable impulse, " Ind. U. Math. J. , 24, 1169-1190.

Eyzaquirre, C. , and S. W. Kuffler, 1955, "Further study of soma, dendrite and axon excitation in single neurons, " J. Gen. Physiol. 39, 121-153.

Fatt, P. , 1957, "Electric potentials occurring around a neurone during its antidromic activation, " J. Neurophysiol. 20, 27-60.

Fender, D. H. and B. Julsez, 1967, "Extension of Panum's fusional area in binocularly stabilized vision, " J. Opt. Soc. Am. 57, 819-830.

Fisher, R. A. , 1937, "The wave of advance of advantageous genes," Ann. Eugen. (now Ann. Human Genet. ) 7 , 355-369.

Fishman, S. N. , B. I. Khodorov and M. V. Volkenshtein, 1972, "Molecular mechanisms of change in the ionic permeability of an electro-excitable membrane - II. Model of the process of activation, " Biophysics 17, 637-643.

FitzHugh, R., 1955, "Mathematical models of threshold phenomena in the nerve membrane," <u>Bull</u>. <u>Math</u>. <u>Biophys.</u> <u>17</u>, 257-278.

FitzHugh, R., 1961, "Impulses and physiological states in theoretical models of nerve membrane," <u>Biophys.</u> <u>J.</u> <u>1</u>, 445-466.

FitzHugh, R., 1962, "Computation of impulse initiation and saltatory conduction in a myelinated nerve fiber," <u>Biophys.</u> <u>J.</u> <u>2</u>, 11-21.

FitzHugh, R., 1965, "A kinetic model of the conductance changes in nerve membrane," <u>J.</u> <u>Cell.</u> <u>Compar.</u> <u>Physiol.</u> <u>66</u>, 111-117.

FitzHugh, R., 1969, "Mathematical models of excitation and propagation in nerve," in H. P. Schwan (ed.), <u>Biological Engineering</u>, McGraw-Hill, 1-85.

FitzHugh, R., 1973, "Dimensional analysis of nerve models," <u>J.</u> <u>Theoret.</u> <u>Biol.</u> <u>40</u>, 517-541.

FitzHugh, R. and H. A. Antosiewicz, 1959, "Automatic computation of nerve excitation-detailed corrections and additions," <u>SIAM</u> <u>J.</u> <u>7</u>, 447-458.

FitzHugh, R. and K. S. Cole, 1973, "Voltage and current clamp transients with membrane dielectric loss," <u>Biophys.</u> <u>J.</u> <u>13</u>, 1125-1140.

Frankel, S., 1955, "On the design of automata and the interpretation of cerebral behavior," <u>Psychometrika</u> <u>20</u>, 149-162.

Frankenhaeuser, B. and A. L. Hodgkin, 1957, "The action of calcium on the electrical properties of squid axons," <u>J.</u> <u>Physiol.</u> <u>137</u>, 218-244.

Frankenhaeuser, B. and A. F. Huxley, 1964, "The action potential in the myelinated nerve fibre of <u>Xenopus</u> <u>Laevis</u> as computed on the basis of voltage clamp data," <u>J.</u> <u>Physiol.</u> <u>171</u>, 302-315.

Fricke, H., 1923, "The electric capacity of cell suspensions," <u>Phys.</u> <u>Rev.</u> <u>21</u>, 708-709.

Fricke, H. , 1925a, 1926, "A mathematical treatment of the electric conductivity and capacity of disperse systems, " Phys. Rev. 24, 575-587; 26, 678-681.

Fricke, H. , 1925b, "The electric capacity of suspensions with special reference to blood, " J. Gen. Physiol. 9, 137-152.

Frölich, H. , 1970, "Long range coherence and the action of enzymes, "Nature 228, 1093.

Furakawa, T. and E. J. Furshpan, 1963, "Two inhibitory mechanisms in the Mauthner neurons of goldfish, " J. Neurophysiol. 26, 140-176.

Furshpan, E. J. and D. D. Potter, 1959a, "Transmission at the giant motor synapses of the crayfish, " J. Physiol. 145, 289-325.

Furshpan, E. J. and D. D. Potter, 1959b, "Slow post-synaptic potentials recorded from the giant motor fibre of the crayfish," J. Physiol. 145, 326-335.

Gabor, D. , 1968a, "Holographic model of temporal recall, " Nature 217, 584.

Gabor, D. , 1968b, "Improved holographic model of temporal recall, " Nature 217, 1288.

Gabor, D. , 1969, "Associative holographic memories, " IBM J. Res. Develop. 13, 156-159.

Gardner, M. F. and J. L. Barnes, 1942, Transients in Linear Systems, Wiley, New York.

Gasser, H. S. and Erlanger, J. , 1922, "A study of the action currents of nerve with a cathode ray oscillograph, " Am. J. Physiol. 62, 496-524.

Geduldig, D. and R. Gruener, 1970, "Voltage clamp of the Aplysia giant neurone: early sodium and calcium currents, " J. Physiol. 211, 217-244.

Gemme, G. , 1969, "Axon membrane crystallites in insect photo-receptors, " in A. Engstrom and B. Strandberg (eds. ) Nobel Symposium 11, Wiley-Interscience, New York, 305.

George, F. H. , 1972, "Mechanism, interrogation and incomplete-ness, " Kybernetes 1, 109-114.

Geselowitz, D. B. , 1966, "Comment on the core conductor model," Biophys. J. 6, 691-692.

Geselowitz, D. B. , 1967, "On bioelectric potentials in an inho-mogeneous volume conductor," Biophys. J. 7, 1-11.

Giaever, I. and K. Megerle, 1962, "The superconductive tunnel junction as an active device, " Trans. IRE Electron. Devices, ED-9, 459-461.

Gödel, K. , 1931, Über formal unentscheidbare  Sätze der Principia Mathematica und verwandter Systeme, " Monatshefte f. Math. u. Physik, 38, 173-198.

Goldman, D. E. , 1943, "Potential, impedance and rectification in membranes, " J. Gen. Physiol. 27, 37-60.

Goldman, D. E. , 1964, "A molecular structural basis for the excita-tion properties of axons, " Biophys. J. 4, 167-188.

Goldman, L. and J. S. Albus, 1968, "Computation of impulse con-duction in myelinated fibers; theoretical basis of the velocity diameter relation, " Biophys. J. 8, 596-607.

Goldstein, S. S. and W. Rall, 1974, "Changes of action potential, shape and velocity for changing core conductor geometry, " Biophys. J. 14, 731-757.

Goldup. A. , S. Ohki and J. F. Danielli, 1970, "Black lipid films," in J. F. Danielli, A. C. Riddiford, and M. D. Rosenberg (eds.), Recent Progress in Surface Science, New York, 193-260.

Good, I. J. , 1965, "Speculations concerning the first ultraintelli-gent machine, " Adv. Computers 6, 31-88.

Gorter, E. and F. Grendel, 1925, "On bimolecular layers of lipoids on the chromocytes of the blood, " J. Exptl. Med. 41, 439-443.

Granit, R. , L. E. Leksell, and C. R. Skoglund, 1944, "Fiber interaction in injured or compressed region of nerve, " Brain 67, 125-140.

Green, D. E. , 1971, "Membrane structure, " Science, 174, 863-867.

Green, D. E. , S. Ji, and R. F. Brucker, 1972, "Structure-function unitization model of biological membranes, " Bioenergetics, 4, 527-558.

Green, R. M. , 1953, Commentary on the Effect of Electricity on Muscular Motion (translation from Luigi Galvani's "De Viribus Electricitatis in Motu Musculari Commentarius"), Elizabeth Licht, Publisher, Cambridge, Mass.

Greenberg, J. M. , 1973, "A note on the Nagumo equation, " Quart. J. Math. 24, 307-314.

Greene, P. H. , 1962, "On looking for neural networks and 'cell assemblies' that underlie behavior, " Bull. Math. Biophys. 24, 247-275; 395-411.

Griffith, J. S. , 1963a, "On the stability of brain-like structures," Biophys. J. 3, 299-308.

Griffith, J. S. , 1963b, "A field theory of neural nets: I. Derivation of field equations, " Bull. Math. Biophys. 25, 111-120.

Griffith, J. S. , 1965, "A field theory of neural nets: II. Properties of field equations, " Bull. Math. Biophys. 27, 187-195.

Griffith, J. S. , 1967, A View of the Brain, Oxford University Press.

Griffith, J. S. , 1971, Mathematical Neurobiology, Academic, New York.

Grossberg, S. , 1969a, "Embedding fields: a theory of learning with physiological implications," J. Math. Psychol. 6, 209-239.

Grossberg, S. , 1969b, "On learning, information, lateral inhibition, and transmitters, " Math. Biosci. 4, 255-310.

Grossberg, S. , 1970, "Neural pattern discrimination, " J. Theoret. Biol. 27, 291-337.

Grossberg, S. , 1971, On the dynamics of operant conditioning, " J. Theoret. Biol. 33, 225-255.

Grossberg, S. , 1972a, "Neural expectation: cerebellar and retinal analogs of cells fired by learnable or unlearned pattern classes, " Kybernetik 10, 49-57.

Grossberg, S. , 1972b, "A neural theory of punishment and avoidance, " Math. Biosci. 15, 39-67; 253-285.

Grossberg, S. , 1973, "Contour enhancement, short term memory, and constancies in reverberating neural networks, " Studies Appl. Math. 52, 213-257.

Grossberg, S. , 1974, "Classical and instrumental learning by neural networks, " Prog. Theoret. Biol. 3, 51-141.

Grossberg, S. , 1975, "A neural model of attention, reinforcement, and discrimination in learning, " C. Pfeiffer (ed. ), Intnat'l Rev. Neurobiol. 18, 263-325.

Grossberg, S. , 1976a, "On the development of feature detectors in the visual cortex with applications to learning and reaction-diffusion systems," Biol. Cybernetics 21, 145-159.

Grossberg, S. , 1976b, "Adaptive pattern classification and universal recoding, I: Parallel development and coding of neural feature detectors, " Biol. Cybernetics 23, 121-134.

Grossberg, S. , 1976c, "Adaptive pattern classification and universal recoding, II: Feedback, expectation, olfaction, and illusions, " Biol. Cybernetics 23, 187-202.

Grossberg, S. and D. Levine, 1975, "Some developmental and attentional biases in the contrast enhancement and short term memory of recurrent neural networks, " J. Theoret. Biol. 53, 341-380.

Grossman, Y. , M. E. Spira, and I, Parnas, 1973, "Differential flow of information into branches of a single axon, " Brain Res., 64, 379-386.

Grundfest, H. , 1958, "Electrophysiology and pharmacology of dendrites, " Electroenceph. Clin. Neurophysiol. Suppl. , 10, 22-41.

Grundfest, H. and J. Magnes, 1951, "Excitability changes in dorsal roots produced by electrotonic effects from adjacent afferent activity, " Am. J. Physiol. 164, 502-508.

Gutman, A. M. , 1971, "Further remarks on the effectiveness of the dendrite synapses, " Biophysics 16, 131-138.

Gutman, A. and A. Shimoliunas, 1973, "Finite dendrite with an N-shaped current-voltage characteristic for the membrane, " Biophysics 18, 1013-1016.

Guttman, R., L. Feldman, and H. Lecar, 1974, "Squid axon membrane response to white noise stimulation, " Biophys. J. 14, 941-955.

Hahn, W. , 1963, Theory and Application of Liapunov's Direct Method, Prentice-Hall, Englewood Cliffs, N. J.

Hall, J. E. , C. A. Mead, and G. Szabo, 1973, "A barrier model for current flow in lipid bilayer membranes, " J. Membrane Biol. 11, 75-97.

Hamlyn, L. H. , 1963, "An electron microscope study of the pyramidal neurons in the Ammon's Horn of the rabbit, " J. Anat. (Lond. ) 97. 189-201.

Harmon, L. D. and E. R. Lewis, 1966, "Neural modeling, " Physiol. Rev. 46, 513-591.

Harth, E. M. , T. J. Csermely, B. Beek, and R. D. Lindsay, 1970, "Brain functions and neural dynamics, " J. Theoret. Biol. 26, 93-120.

Harth, E. M. and S. L. Edgar, 1967, "Association by synaptic facilitation in highly damped neural nets, " Biophys. J. 7, 689-717.

Hastings, S. P. , 1972, "On a third order differential equation from biology, " Quart. J. Math. 23, 435-448.

Hastings, S. P. , 1974, "The existence of periodic solutions to Nagumo's equation, " Quart. J. Math. 25, 369-378.

Hastings, S. P. , 1975, "Some mathematical problems from neurobiology, " Am. Math. Month. 82, 881-895.

Hastings, S. P. , 1976a, "On travelling wave solutions of the Hodgkin-Huxley equations, " Arch. Rat. Mech. Anal. 60, 229-257.

Hastings, S. P. , 1976b, "The existence of homoclinic and periodic orbits for the FitzHugh-Nagumo equations, " Quart. J. Math. 27, 123-134.

Hatase, O. , T. Wakabayashi, H. Hayashi, and D. E. Green, 1972, "Collapse and extension of the headpiece-stalk projections in mitochondrial electron transport particles, " Bioenergetics 3, 509-514.

Hebb, D. O. , 1949, Organization of Behavior, Wiley, New York.

Hebb, D. O. , 1972, Textbook of Psychology (3rd. ed. ), Saunders, Philadelphia.

Hellerstein, D. , 1968, "A generalization of the theory of electrotonus, " Biophys. J. 8, 358-379.

Helmholtz, H. , 1850, "Messungen über den zeitlichen Verlauf der Zuchung animalischer Muskeln und die Fortpflanzungsgeschwindigkeit der Reizung in den Nerven, " Arch. Anat. Physiol. , 276-364.

Hendrix, C. E. , 1965, "Transmission of electric fields in cortical tissue: a model for the origin of the alpha rhythm, " Bull. Math. Biophys. 27, 197-213.

Hering, E., 1882, "Beiträge zur allgemeinen Nerven- und Muskel-physiologie. IV. Über Nervenreizung durch den Nervenstrom" Sitzungsber. k. Akad. Wiss. 85, pt. 3, 237-275.

Hermann, L., 1879, Hanbuch der Physiologie: (a) Bewegungsap-parate; (b) Nervensystems, F. C. W. Vogel, Leipzig.

Hermann, L., 1905, "Beiträge zur Physiologie und Physik des Nerven," Arch. ges. Physiol. 109, 95-144.

Hild, W., and I. Tasaki, 1962, "Morphological and physiological properties of neurons and glial cells in tissue cluture," J. Neurophysiol. 25, 277-304.

Hille, B., 1970, "Ionic channels in nerve membranes," Prog. Biophys. Molec. Biol. 21, 3-32.

Hodgkin, A. L., 1951, "The ionic basis of electrical activity in nerve and muscle," Biol. Rev. 26, 339-409.

Hodgkin, A. L., 1954, "A note on conduction velocity," J. Physiol. 125, 221-224.

Hodgkin, A. L., 1964, The Conduction of the Nervous Impulse. Liverpool University Press.

Hodgkin, A. L. and A. F. Huxley, 1952a, "Currents carried by sodium and potassium ions through the membrane of the giant axon of Loligo," J. Physiol. 116, 449-472.

Hodgkin, A. L. and A. F. Huxley, 1952b, "The components of membrane conductance in the giant axon of Loligo," J. Physiol. 116, 473-496.

Hodgkin, A. L. and A. F. Huxley, 1952c, "The dual effect of mem-brane potential on sodium conductance in the giant axon of Loligo," J. Physiol. 116, 497-506.

Hodgkin, A. L. and A. F. Huxley, 1952d, "A quantitative descrip-tion of membrane current and its application to conduction and excitation in nerve," J. Physiol. 117, 500-544.

Hodgkin, A. L. , A. F. Huxley, and B. Katz, 1952, "Measurement of current-voltage relations in the membrane of the giant axon of Loligo, " J. Physiol. 116, 424-448.

Hodgkin, A. L. and B. Katz, 1949, "The effect of sodium ions on the electrical activity of the giant axon of the squid, " J. Physiol. 108, 37-77.

Hoorweg, J. L. , 1898, "Üeber die elektrischen Eigenschaften der Nerven " Arch. ges. Physiol. 71, 128-157.

Howard, R. E. and R. M. Burton, 1968, "Thin lipid membranes with aqueous interfaces: apparatus design and methods of study, " J. Am. Oil Chem. Soc. 45, 202-229.

Hoyt, R. C. , 1963, "The squid giant axon; mathematical models," Biophys. J. 3, 399-431.

Hoyt, R. C. , 1968, "Sodium inactivation in nerve fibers, " Biophys. J. 8, 1074-1097.

Hoyt, R. C. and W. J. Adelman, Jr. , 1970, "Sodium inactivation; experimental test of two models, " Biophys. J. 10, 610-617.

Hoyt, R. C. and J. D. Strieb, 1971, "A stored charge model for the sodium channel, " Biophys. J. 11, 868-885.

Hubel, D. H. and R. M. Wiesel, 1962, "Receptive fields, binocular interaction and functional architecture in the cat's visual cortex, " J. Physiol. 160, 106-154.

Hurewicz, W. , 1958, Lectures on Ordinary Differential Equations, Wiley, New York.

Huxley, A. F. , 1959a, "Ion movements during nerve activity, " Ann. N. Y. Acad. Sci. 81, 221-246.

Huxley, A. F. , 1959b, "Can a nerve propagate a subthreshold disturbance? " J. Physiol. 148, 80P - 81P.

Il'inova, T. M. and R. V. Khokhlov, 1963, "Wave processes in lines with nonlinear shunt resistance, " <u>Rad</u>. <u>Eng</u>. <u>Elec</u>. <u>Phys</u>. <u>8</u>, 1864-1872.

Inoue, I. , Y. Kobatake, and I. Tasaki, 1973, "Excitability, in-stability and phase transitions in squid axon membrane under internal perfusion with dilute salt solutions, " <u>Biochimica</u> <u>et</u> <u>Biophysica</u> <u>Acta</u> 307, 471-477.

Inoue, I. , I. Tasaki, and Y. Kobatake, 1974, "A study of the ef-fects of externally applied sodium-ions and detection of spa-tial non-uniformity of the squid axon membrane under internal perfusion, " <u>Biophys</u>. <u>Chem</u>. 2, 116-126.

Isaacs, C. D. , 1970, "Analog-digital-hybrid studies of the re-formulated equations of Hodgkin-Huxley, " <u>Math</u>. <u>Biosci</u>. <u>7</u>, 305-312.

Ishihara, T. , 1971a, "Local reverberations in the nervous system and conditioned reflex, " <u>Math</u>. <u>Biosci</u>. <u>12</u>, 23-31.

Ishihara, T. , 1971b, "Local reverberations in the nervous system and memory, " <u>Math</u>. <u>Biosci</u>. <u>12</u>, 225-233.

Ishihara, T. and M. Sato, 1974, "Variation and stability of re-verberations in threshold systems, " <u>Math</u>. <u>Jap</u>. <u>19</u>, 357-369.

Jack, J. J. B. , S. Miller, R. Porter, and S. J. Redman, 1971, "The time course of minimal excitatory post-synaptic poten-tials evoked in spinal motoneurons by group Ia afferent fibres, " <u>J</u>. <u>Physiol</u>. <u>215</u>, 353-380.

Jack, J. J. B. , D. Noble, and R. W. Tsien, 1975, <u>Electric</u> <u>Current</u> <u>Flow</u> <u>in</u> <u>Excitable</u> <u>Cells</u>, Clarendon Press, Oxford.

Jack, J. J. B. and S. J. Redman, 1971a, "The propagation of transient potentials in some linear cable structures, " <u>J</u>. <u>Physiol</u>. <u>215</u>, 283-320.

Jack, J. J. B. and S. J. Redman, 1971b, "An electrical description of the motoneuron, and its application to the analysis of syn-aptic potentials, " <u>J</u>. <u>Physiol</u>. <u>215</u>, 321-352.

Jain, M. K. , 1972, <u>The</u> <u>Bimolecular</u> <u>Lipid</u> <u>Membrane</u>, Van Nost-rand Reinhold, New York.

Jain, M. K. , R. H. L. Marks, and E. H. Cordes, 1970, "Kinetic model of conductance changes across excitable membranes, " Proc. N. A. S. 67, 799-806.

Jakobsson, E. , 1973, "The physical interpretation of mathematical models for sodium permeability changes in excitable membranes, " Biophys. J. 13, 1200-1211.

Jakobsson, E. and C. Scudiero, 1975, "A transient excited state model for sodium permeability changes in excitable membranes, " (in preparation).

Jasper, H. H. and A. M. Monnier, 1938, "Transmission of excitation between excised non-myelinated nerves. An artificial synapse, " J. Cell. Compar. Physiol. 11, 259-277.

Johnson, W. J. , 1968, "Nonlinear wave propagation on superconducting tunneling junctions, " Ph. D. thesis, University of Wisconsin.

Kanel',Ya. I. , 1962, "On the stability of the Cauchy problem for equations occurring in the theory of flames, " Mathematicheskii Sbornik, 59 (101), 246-288.

Kaplan, S. , and D. Trujillo, 1970, "Numerical studies of the partial differential equations governing nerve impulse conduction: The effect of Lieberstein's inductance term, " Math. Biosci. 7, 379-404.

Kato, G. , 1924, The Theory of Decrementless Conduction in Narcotized Region of Nerve, Nankōdō, Tokyo, Japan.

Kato, G. , 1934, Microphysiology of Nerve, Maruzen, Tokyo, Japan.

Kato, G. , 1970, "The road a scientist followed, " Ann. Rev. Physiol. , 32, 1-20.

Katz, B. , 1966, Nerve, Muscle and Synapse, McGraw-Hill, New York. Reference in chapters on membrane structure and function.

Katz, B. and O. H. Schmitt, 1939, "Excitability changes in a nerve fiber during the passage of an impulse in an adjacent fiber," J. Physiol. 96, 9P-10P.

Katz, B. , and O. H. Schmitt, 1940, "Electric interaction between two adjacent nerve fibers," J. Physiol. 97, 471-488.

Katz, B. , and O. H. Schmitt, 1942, "A note on the interaction between nerve fibers," J. Physiol. , 100, 369-371.

Kelvin, Lord (William Thompson), 1855, "On the theory of the electric telegraph," Proc. Roy. Soc. (Lond. ) 7, 382-399.

Khodorov, B. I. , 1974, The Problem of Excitability, Plenum, New York.

Khodorov, B. I. and Ye. N. Timin, 1970, "III. Transformation of rhythms in the cooled part of the fibre," Biophysics, 15, 526-536.

Khodorov, B. I. and Ye. N. Timin, 1971, "V. Phenomena of Vvedenskii in portions of the fiber with reduced sodium and potassium conductivity of the membrane," Biophysics, 16, 513-523.

Khodorov, B. I. , Ye. N. Timin, N. V. Pozin, and L. A. Shmelev, 1971, "IV. Conduction of a series of impulses through a portion of the fibre with increased diameter," Biophysics, 16, 96-104.

Khodorov, B. I. , Ye. N. Timin, S. Ya. Vilenkin, and F. B. Gul'ko, 1969, "Theoretical analysis of the mechanisms of conduction of a nerve pulse over an inhomogeneous axon. I. Conduction through a portion with increased diameter," Biophysics, 14, 323-335.

Khodorov, B. I. , Ye. N. Timin, S. Ya. Vilenkin and F. B. Gul'ko, 1970, "II. Conduction of a single impulse across a region of the fiber with modified functional properties," Biophysics 15, 145-152.

Kilkson, R. , 1969, "Symmetry and function of biological systems at the macromolecular level, " in A. Engström and B. Strandberg (eds. ), Nobel Symposium 11, Wiley-Interscience, New York, 257-266.

Kishimoto, U. , 1965, "Voltage clamp and internal perfusion studies on Nitella internodes, " J. Cell. Compar. Physiol. 66 Suppl. 2, 43-53.

Kitagawa, T. , 1973, "Dynamical systems and operators associated with a single neuronic equation, " Math. Biosci. 18, 191-244.

Knight, B. W. , and G. A. Peterson, 1967, "Theory of the Gunn effect, " Phys. Rev. 155, 393-404.

Koestler, A. , 1971, The Ghost in the Machine, Henry Regenery Company, Chicago.

Kolmogoroff, A. , I. Petrovsky, and N. Piscounoff, 1937, "Étude de l'équation de la diffusion avec croissance de la quantite de matière et son application a un problème biologique, " Bull. Univ. Moscow, Série Int. , Al, 1-25.

Kompaneyets, A. S. 1971, "Influence of ohmic resistance of the nerve fiber membrane on certain effects of nervous excitation," Biophysics 16, 926-934.

Kompaneyets, A. S. and V. Ts. Gurovich, 1966, "Propagation of an impulse in a nerve fiber, " Biophysics 11, 1049-1052.

Korn, H. and M. V. L. Bennett, 1971, "Dendritic and somatic impulse  initiation in fish oculomotor neurons during vestibular nystagmus, Brain Res. 27, 169-175.

Korn, H. and M. V. L. Bennett, 1972, "Electrotonic coupling between teleost oculomotor neurons; restriction to somatic regions and relation to function of somatic and dendritic sites of impulse initiation, " Brain Res. 38, 433-439.

Korteweg, D. J. and G. de Vries, 1895, "On the change of form of long waves advancing in a rectangular canal, and on a new type of long stationary wave, " Phil. Mag. 39, 422-443.

Krinskii, V. I. , and Yu. M. Kokoz, 1973, "Analysis of equations of excitable membranes-- I.  Reduction of the Hodgkin-Huxley equations to a second order system, " Biophysics, 18, 533-539.

Krnjević, K. , and R. Miledi, 1959, "Presynaptic failure of neuro-muscular propagation in rats, " J. Physiol. 149, 1-22.

Kryloff, N. and N. Bogoliuboff, 1947, Introduction to Nonlinear Mechanics, Princeton University Press.

Kryshtal', O. A. , I. S. Magura, and N. T. Parkhomenko, 1969, "Ionic currents across the membrane of the some of giant neurones of the edible snail with clamped shifts of the membrane potential, " Biophysics 14, 987-989.

Kuhn, T. S. , 1962, The Structure of Scientific Revolutions, University of Chicago Press.

Kuno, M. and J. T. Miyahara, 1969, "Non-linear summation of unit synaptic potentials in spinal motoneurons of the cat, " J. Physiol. 201, 465-477.

Kunov, H. , 1965, "Controllable piecewise-linear lumped neuristor realization, " Electron. Lett. 1, 134.

Kunov, H. , 1966, "Nonlinear transmission lines simulating nerve axon, " Thesis, Electronics Laboratory, Technical University of Denmark.

Kunov, H. , 1967, "On recovery in a certain class of neuristors, " Proc. IEEE, 55, 428-429.

Landowne, D. , 1972, "A new explanation of the ionic currents which flow during the nerve impulse, " J. Physiol. 222, 46P-47P.

Landowne, D. , 1973, "Movement of sodium ions associated with the nerve impulse, " Nature, 242, 457-459.

Langmuir, N. , 1917, "The constitution and fundamental properties of solids and liquids, " J. Am. Chem. Soc. 39, 1848-1906.

Langmuir, I. , and D. F. Waugh, 1938, "The absorption of proteins at oil-water interfaces and artificial protein-lipoid membranes, " J. Gen. Physiol. 21, 745-755.

Lawrence, A. S. C. , 1929, Soap Films, Bell, London.

Lefschetz, S. , 1962, Differential Equations: Geometric Theory, Wiley-Interscience, New York.

Legéndy, C. R. , 1967, "On the scheme by which the human brain stores information, " Math. Biosci. 1, 555-597.

Lehninger, A. L. , 1968, "The neuronal membrane, " Proc. N. A. S. 60, 1069-1080.

Lehninger, A. L. , 1970, Biochemestry, Worth Pub. Co. , New York.

Leibovic, K. N. , 1972, Nervous System Theory, Academic, New York.

Leibovic, K. N. , and N. H. Sabah, 1969, "On synaptic transmission, neural signals and psychophysiological phenomena, " in Information Processing in the Nervous System (K. N. Leibovic (ed. ), Springer-Berlag, New York.

Levine, S. V. , D. L. Rozental' and Ya. Yu. Komissarchik, 1968, "Structural changes in the axon membrane on excitation, " Biophysics 13, 214-217.

Levine, D. S. and S. Grossberg, 1976, "Visual illusions in neural networks: line neutralization, tilt after effect, and angle expansion, " J. Theoret. Biol. 61, 477-504.

Lieberstein, H. M. , 1967a, "On the Hodgkin-Huxley partial differential equation, " Math. Biosci. 1, 45-69.

Lieberstein, H. M. , 1967b, "Numerical studies of the steady-state equations for a Hodgkin-Huxley model, " Math. Biosci. 1, 181-211.

Lieberstein, H. M. , 1973, Mathematical Physiology: Blood Flow and Electrically Active Cells, Am. Elsevier, New York.

Lieberstein, H. M. and M. A. Mahrous, 1970, "A source of large inductance and concentrated moving magnetic fields on axons," Math. Biosci, 7, 41-60.

Lillie, R. S., 1925, "Factors affecting transmission and recovery in the passive iron nerve model, " J. Gen. Physiol. 7, 473-507.

Lillie, R. S., 1936, "The passive iron wire model of protoplasmic and nervous transmission and its physiological analogues, " Biol. Rev. 11, 181-209.

Lin, J. and P. B. Kahn, 1976, "Averaging methods in predator-prey systems and related biological models, " J. Theoret. Biol. 57, 73-102.

Lindgren, A. G. and R. J. Buratti, 1969, "Stability of waveforms on active non-linear transmission lines, " Trans. IEEE Circuit Theory, CT-16, 274-279.

Llinás, R., and C. Nicholson, 1971, "Electrophysiological properties of dendrites and somata in alligator Purkinje cells, " J. Neurophysiol. 34, 532-551.

Llinás, R., C. Nicholson, J. A. Freeman, and D. E. Hillman, 1968, "Dendritic spikes and their inhibition in alligator Purkinje cells, " Science 160, 1132-1135.

Llinás, R., C. Nicholson, and W. Precht, 1969, "Preferred centripital conduction of dendrite spikes in alligator Purkinje cells, " Science 163, 184-187.

Longuet-Higgins, H. C., 1968, "Holographic model of temporal recall, " Nature 217, 104.

Lopes de Silva, F. H., A. Hocks, H. Smits, and L. H. Zetterberg, 1974, "Model of brain rhythmic activity, the alpha-rhythm of the thalamus, " Kybernetik 15, 27-37.

Lorente de Nó, 1947a, "A study of nerve physiology, " Stud. Rockefeller Inst., 132, 384-476.

Lorente de Nó, R., 1947b, "Action potential of the motoneurons of the hypoglossus nucleus," J. Cell. Compar. Physiol. 29, 207-287.

Lorente de Nó, R., 1960, "Decremental conduction and summation of stimuli delivered to neurons at distant synapses," in Structure and Function of the Cerebral Cortex, (D. B. Tower and J. P. Schade, eds.), American Elsevier, New York, 278-281.

Lorente de Nó, R., and G. A. Condouris, 1959, "Decremental conduction in peripheral nerve. Integration of stimuli in the neuron," Proc. N. A. S. 45, 592-617.

Lucas, K., 1909, "The 'all-or-none' contraction of amphibian skeletal muscle," J. Physiol. 38, 113-133.

Luk'yanov, A. S., 1970, "Generation of action potentials by dendrites in the frog optic tectum," Fisiologicheskii Zhurnal SSSR 56, 1130-1135; translated in Neurosci. Transl. 16, 74-78.

Mackey, M. C., 1975, Ion Transport through Biological Membranes, Lecture Notes in Biomathematics, Springer-Verlag, New York.

Maginu, K., 1971, "On asymptotic stability of waveforms on a bistable transmission line," IECE Professional Group on Nonlinear Problems NLP 70-24 (in Japanese).

Markin, V. S., 1970a, "Electrical interaction of parallel nonmyelinated nerve fibres. I. Change in excitability of the adjacent fibre," Biophysics 15, 122-133.

Markin, V. S., 1970b, "II. Collective conduction of impulses," Biophysics 15, 713-721.

Markin, V. S., 1973a, "III. Interaction in bundles," Biophysics 18, 324-332.

Markin, V. S., 1973b, "IV. Role of anatomical inhomogeneities of nerve trunks," Biophysics 18, 539-547.

Markin, V. S. and Yu. A. Chizmadzhev, 1967, "On the propagation of an excitation for one model of a nerve fiber, " Biophysics 12, 1032-1040.

Markin, V. S. and Yu. A. Chizmadzhev, 1972, "Properties of a multicomponent excitable medium, " J. Theoret. Biol. 36, 61-80.

Markin, V. S. and Yu. A. Chizmadzhev, 1974, Indutsirovannui Ionnui Transport (induced ionic transport), Izdatelbstvo "Nauka, " Moscow (in Russian).

Markin, V. S. , P. A. Grigor'ev, and L. N. Yermishkin, 1971, "Forward passage of ions across lipid membranes. I. Mathematical model, " Biophysics 16, 1050-1059.

Markin, V. S. and V. F. Pastushenko, 1969, "Spread of excitation in a model of an inhomogeneous nerve fibre. I. Slight change in dimensions of fibre, " Biophysics 14, 335-344.

Markin, V. S. and V. F. Pastushenko, 1973, "Excitation propagation in septated axons, " T. I. T. J. Life Sci. 3, 87-94.

Marmont, G. , 1949, "Studies on the axon membrane. I. A new method. " J. Cell. Compar. Physiol. 34, 351-382.

Marr, D. , 1969, "A theory of cerebellar cortex, " J. Physiol. , 202, 437-470.

Marr, D. , 1970, "A theory for the cerebral neocortex, " Proc. Roy. Soc. (Lond. ) 176 B, 161-234.

Marrazzi, A. S. and R. Lorente de Nó, 1944, "Interaction of neighboring fibers in myelinated nerve, " J. Neurophys. 7, 83-101.

Maslow, A. H. , 1966, The Psychology of Science, Harper and Row, New York.

Maslow, A. H. , 1968, Toward a Psychology of Being, Van Nostrand, New York.

Mauro, A., 1961, "Anomalous impedance, a phenomenological property of time-variant resistance," Biophys. J., 1, 353-372.

Mauro, A., F. Conti, F. Dodge, and R. Schor, 1970, "Subthreshold behavior and phenomenological impedance of the squid giant axon," J. Gen. Physiol., 55, 497-523.

Mauro, A., A. R. Freeman, J. W. Cooley, and A. Cass, 1972, "Propagated subthreshold oscillatory response and classical electronic response of squid giant axon," Biophysik, 8, 118-132.

McCulloch, W. S. and W. H. Pitts, 1943, "A logical calculus of the ideas immanent in nervous activity," Bull. Math. Biophys. 5, 115-133.

McCulloch, W. S., 1965, Embodiments of Mind, M. I. T. Press, Cambridge, Mass.

McKean, H. P., Jr., 1970, "Nagumo's equation," Adv. Math. 4, 209-223.

McLennan, H., 1970, Synaptic Transmission (2nd ed.), Saunders, Philadelphia.

Mehra, J., 1973, "Quantum mechanics and the explanation of life," Am. Sci. 61, 722-728.

Milner, P. M., 1957, "The cell assembly: Mark II," Psychol. Rev. 64, 242-252.

Minor, A. V. and V. V. Maksimov, 1969, "Passive electrical properties of the model of a flat cell," Biophysics 14, 349-357.

Minsky, M. and S. Papert, 1969, Perceptrons, M. I. T. Press, Cambridge, Mass.

M. I. T. Elec. Eng. Staff, 1943, Applied Electronics, Wiley, New York, 114-124.

Moore, J. W., 1968, "Specifications for nerve membrane models," Proc. IEEE 56, 895-905.

Moore, L. E. and E. Jakobsson, 1971, "Interpretation of the sodium permeability changes of myelinated nerve in terms of linear relaxation theory, " J. Theoret. Biol 33, 77-89.

Moore, J. W. , F. Ramon, R. W. Joyner, and N. Anderson, 1975, "Axon voltage-clamp simulations, " Biophys. J. 15, 11-69.

Morse, P. M. and H. Feshbach, 1953, Methods of Theoretical Physics, McGraw-Hill, New York.

Mountcastle, V. B. , 1957, "Modality and topographic properties of single neurons of cat's somatic sensory cortex, " J. Neurophysiol. 20, 408-434.

Mueller, P. and D. O. Rudin, 1968a, "Resting and action potentials in experimental bimolecular lipid membranes, " J. Theoret. Biol. 18, 222-258.

Mueller, P. and D. O. Rudin, 1968b, "Action potentials induced in bimolecular lipid membranes, " Nature 217, 713-719.

Mueller, P. , D. O. Rudin, H. T. Tien, and W. C. Wescott, 1962, "Reconstitution of cell membrane structure in vitro and its transformation into an excitable system, " Nature 194, 979-980.

Mullins, L. J. , 1959, "An analysis of conductance changes in squid axon, " J. Gen. Physiol. 42, 1013-1035.

Mysels, K. J. , K. Shinoda and S. Frankel, 1959, Soap Films, Studies of Their Thinning and a Bibliography, Pergamon, New York.

Nachmansohn, D. , 1959, Chemical and Molecular Basis of Nerve Activity, Academic, New York.

Nachmansohn, D. and Neuman, E. , 1974, "Properties and function of proteins in excitable membranes: An integral model of nerve excitability, " Ann. N. Y. Acad. Sci. , 227, 275-284.

Nagano, T. , S. Ohteru, and T. Kato, 1967, "Learning machine with nonlinear weighting elements," Electron. Commun. Jap. , 50, 1-7.

Nagel, E. and J. R. Newman, 1958, Gödel's Proof, New York University Press.

Nagumo, J. , S. Arimoto, and S. Yoshizawa, 1962, "An active pulse transmission line simulating nerve axon," Proc. IRE, 50, 2061-2070.

Nagumo, J. , S. Yoshizawa, and Arimoto, 1965, "Bistable transmission lines, " Trans. IEEE Circuit Theory, CT-12, 400-412.

Nakajima, K. , Y. Onodera, T. Nakamura, and R. Sato, 1974, "Analysis of vortex motions on Josephson line, " J. Appl. Phys. 45, 4095-4099.

Nakajima, K. , Y. Onodera, and Y. Ogawa, 1976, "Logic design of Josephson network, " J. Appl. Phys. 47, 1620-1627.

Nakajima, K. , T. Yamashita and Y. Onodera, 1974, "Mechanical analog of active Josephson transmission line, " J. Appl. Phys. 45, 3141-3145.

Namerow, N. S. and J. J. Kappl, 1969, "Conduction in demyelinated axons - a simplified model, " Bull. Math. Bioph. 31, 9-23.

Nernst, W. , 1888, "Zur kinetik der in Lösung befindlichen Körper: Theorie der Diffusion, " Zeit Physik Chem. 2, 613-637.

Nernst, W. , 1889, "Die elektromotorische Wirksamkeit der Ionen," Zeit. Physik. Chem. , 4, 129-181.

Neumke, B. , and P. Läuger, 1969, "Nonlinear electrical effects in lipid bilayer membranes, II.  Integration of the generalized Nernst-Planck equations, " Biophys. J. 9, 1160-1170.

Newton, I. , 1952, Optiks (based on the fourth edition, London, 1730), Dover Publications, New York.

316    References

Nilsson, N. J., 1965, Learning Machines, McGraw-Hill, New York.

Noble, D. , 1962, "A modification of the Hodgkin-Huxley equations applicable to Purkinje fibre action and pacemaker potentials," J. Physiol. 160, 317-352.

Noble, D. , 1966, "Applications of Hodgkin-Huxley equations to excitable tissues, " Physiol. Rev. 46, 1-50.

Noble, D. and R. B. Stein, 1966, "The threshold conditions for initiation of action potentials by excitable cells, " J. Physiol. 187, 129-142.

Noguchi, S. , Y. Kumagai, and J. Oizumi, 1963, "General considerations of the neuristor circuits, " Sci. Rept. Res. Inst. Tohoku U. 14, 155-184.

Nuñez, P. L. , 1974a, "The brain wave equation: a model for the EEG, " Math. Biosci. 21, 279-297.

Nuñez , P. L. , 1974b, "Wave like properties of the alpha rhythm," Trans. IEEE Biomed. Eng. BME-21, 473-482.

Nystrom, R. A. , 1973, Membrane Physiology, Prentice-Hall, Englewood Cliffs, N. J.

Offner, F. F. , 1970, "Kinetics of excitable membranes. Voltage amplification in a diffusion regime, " J. Gen. Physiol. 56, 272-296.

Offner, F. F. , 1971, "Nernst-Planck-Poisson diffusion equation: numerical solution of the boundary value problem, " J. Theoret. Biol. 31, 215-227.

Offner, F. F. , 1972, "The excitable membrane. A physiochemical model. " Biophys. J. 12, 1583-1629.

Offner, F. F. , 1974, "Solution of the time-dependent ionic diffusion equation, " J. Theoret. Biol. 45, 81-91.

Offner, F. F. , Weinberg, A. , and Young, G. , 1940, "Nerve conduction theory: some mathematical consequences of Bernstein's model, " Bull. Math. Biophys. 2, 89-103.

Opatowski, I. , 1950, "The velocity of conduction in nerve fiber and its electric characteristics, " Bull. Math. Biophys. 12, 277-302.

Osterhout, W. J. V. , and S. E. Hill, 1930, "Salt bridges and negative variations, " J. Gen. Physiol. 13, 547-552.

Otani, T. , 1937, "Über eine Art Hemmung und Bahnung in Folge der Wechselbeziehungen Nervenfasern zueinander, " Japan. J. Med. Sci. (III Biophysics) 4, 355-372.

Parmentier, R. D. , 1967, "Stability analysis of neuristor waveforms, " Proc. IEEE 55, 1498-1499.

Parmentier, R. D. , 1968, "Neuristor waveform stability analysis by Lyapunov's second method, " Proc. IEEE 56, 1607-1608.

Paramentier, R. D. , 1969, "Recoverable neuristor propagation on superconductive tunnel junction strip lines, " Solid-State Electron. 12, 287-297.

Parmentier, R. D. , 1970a, "Neuristor analysis techniques for non-linear distributed electronic systems, " Proc. IEEE 58, 1829-1837.

Parmentier, R. D. , 1970b, "Estimating neuristor excitation thresholds, " Proc. IEEE 58, 605-606.

Parnas, I. , 1972, "Differential block at high frequency of branches of a single axon innervating two muscles, " J. Neurophysiol. 35, 903-914.

Pastushenko, V. F. , 1975, "The wave regime of the activity of the neuronal network, " Biophysics 20, 503-511.

Pastushenko, V. F. , Yu. A. Chizmadzhev, and V. S. Markin, 1975, "Speed of excitation in the reduced Hodgkin-Huxley model-- a. Rapid relaxation of the sodium current, " Biophysics 20, 685-692; "b. Slow relaxation of sodium current, " ibid. , 894-901.

Pastushenko, V. F. and V. S. Markin, 1969, "Propagation of excitation in a model of an inhomogeneous nerve fibre. II. Attenuation of pulse in the inhomogeneity, " Biophysics 14, 548-552.

Pastushenko, V. F. , and V. S. Markin, 1973, "Spread of excitation in a nerve fiber with septa - II.   Blocking of the impulse by the septum, " Biophysics 18, 740-754.

Pastushenko, V. F. , V. S. Markin, and Yu. A. Chizmadzhev, 1969a, "Propagation of excitation in a model of the inhomogeneous nerve fibre. III.   Interaction of pulses in the region of the branching node of a nerve fibre, " Biophysics 14, 929-937.

Pastushenko, V. F. , V. S. Markin, and Yu. A. Chizmadzhev, 1969b, "IV.   Branching as a summator of nerve pulses, " Biophysics 14, 1130-1138.

Patlak, C. S. , 1955, "Potential and current distribution in nerve: the effect of the nerve sheath, the number of fibers, and the frequency of alternating current stimulation, " Bull. Math. Biophys. 17, 287-307.

Penfield, W. and P. Perot, 1963, "The brain's record of auditory and visual experience-- a final summary and discussion, " Brain 86, 595-696.

Pickard, W. F. , 1966, "On the propagation of the nervous impulse down medullated and unmedullated fibers, " J. Theoret. Biol. 11, 30-45.

Pickard, W. F. , 1968, "A contribution to the electromagnetic theory of the unmyelinated axon, " Math. Biosci. 2, 111-121.

Pickard, W. F. , 1969a, "The electromagnetic theory of electrotonus along an unmyelinated axon, " Math. Biosci. 5, 471-494.

Pickard, W. F. , 1969b, "Estimating the velocity of propagation along myelinated and unmyelinated fibers, " Math. Biosci. 5, 305-319.

Pickard, W. F. , 1974, "Electrotonus on a nonlinear dendrite, " Math. Biosci. 20, 75-84.

Planck, M. , 1890a, "Ueber die Erregung von Elektricität und Wärme in Elektrolyten, " Ann. Physik. Chem. 39, 161-186.

Planck, M. , 1890b, "Ueber die Potential differenz zwischen zwei verdünnten Lösungen binärer Elektrolyte, " Ann. Physik. Chem. 40, 561-576.

Platt, J. R. , 1966, The Step to Man, Wiley, New York, pp. 141-155.

Plonsey, R. , 1964, "Volume conductor fields of action currents, " Biophys. J. , 4, 317-328.

Plonsey, R. , 1965, "An extension of the solid angle potential formulation for an active cell, " Biophys. J. , 5, 663-667.

Pokrovskii, A. N. , 1970, "Mechanism of origin of electric potentials in nerve tissue, " Biophysics 15, 914-921.

Polakov, I. V. , 1973, "Rising, quasi-stationary voltage waves in nonhomogeneous lines with nonlinear resistance, " Rad. Eng. Elec. Phys. 18, 722-726.

Polanyi, M. 1962, "Tacit knowing : its bearing on some problems of philosophy, " Rev. Mod. Phys. 34, 601-616.

Polanyi, M. , 1965, "The structure of consciousness, " Brain 88, 799-810.

Poritsky, R. , 1969, "Two and three dimensional ultrastructure of boutons and glial cells on the motoneuronal surface in the cat spinal cord, " J. Compar. Neur. 135, 423-452.

Predonzani, G. , and A. Roveri, "Criteri di stabilità per una linea contenente bipoli attive concentrati non-lineari, " Note, Recensioni e Notizie, 17, 1433-1453.

Pribram, K. H. , 1969, "The neurophysiology of remembering" Sci. Am. (Jan. ), 73-85.

Pribram, K. H. , 1971, Languages of the Brain, Prentice-Hall, Englewood Cliffs, N. J.

Purpura, D. P. , and H. Grundfest, 1956, "Nature of dendritic potentials and synaptic mechanisms in cerebral cortex of cat," J. Neurophysiol. , 19, 573-595.

Rall, W. , 1959, "Branching dendritic trees and motoneuron membrane resistivity, " Exptl. Neurol. 1, 491-527.

Rall, W. , 1962a, "Theory of physiological properties of dendrites," Ann. N. Y. Acad. Sci. , 96, 1071-1092.

Rall, W. , 1962b, "Electrophysiology of a dendritic neuron model," Biophys. J. , 2, 145-167.

Rall, W. , 1964, "Theoretical significance of dendritic trees for neuronal input-output relations, " in R. F. Reiss (ed. ), Neural Theory and Modeling, Stanford University Press, 73-97.

Rall, W. , 1967, "Distinguishing theoretical synaptic potentials computed for different soma-dendritic distributions of synaptic input, " J. Neurophysiol. 30, 1138-1168.

Rall, W. , 1969, "Distributions of potential in cylindrical coordinates and time constants for a membrane cylinder, " Biophys. J. 9, 1509-1541.

Rall, W. , 1970, "Dendritic neuron theory and dendrodendritic synapses in a simple cortical system, " in F. O. Schmitt (ed. ), The Neurosciences: Second Study Program, The Rockefeller University Press, New York.

Rall, W. and J. Rinzel, 1973, "Branch input resistance and steady attenuation for input to one branch of a dendritic neuron model," Biophys. J. 13, 648-688.

Rall, W. , and G. M. Shepherd, 1968, "Theoretical reconstruction of field potentials and dendrodendritic synaptic interactions in olfactory bulb, " J. Neurophysiol. 31, 884-915.

Ramo, S. and J. R. Whinnery, 1953, Fields and Waves in Modern Radio, Wiley, New York.

Ramón y Cajal, S. , 1908, "Structure et connexions des neurons," Archivio di Fisiologia 5, 1-25.

Ramón y Cajal, S. , 1952, Histologie du Système Nerveux, Cons. Sup. de Invest. Cientificas, Madrid.

Ramón y Cajal, S. , 1954, Neuron Theory or Reticular Theory, Cons. Sup. de Invest. Cientificas, Madrid.

Ramon-Moliner, E. , 1962, "An attempt at classifying nerve cells on the basis of their dendritic patterns, " J. Compar. Neurol. 119, 211-227.

Rapoport, A. , 1952, "'Ignition' phenomena in random nets, " Bull. Math. Biophys. 14, 35-44.

Reible, S. A. and A. C. Scott, 1975, "Pulse propagation on a superconductive neuristor, " J. Appl. Phys. 46, 4935-4945.

Renshaw, B. and P. O. Therman, 1941, "Excitation of entraspinal mammalian axons by nerve impulses in adjacent axons, " Am. J. Physiol. 133, 96-105.

Revenko, S. V. , Y. E. Timin, and B. I. Khodorov, 1973, "Special features of the conduction of nerve impulses from the myelinized part of the axon into the non-myelinated terminal, " Biophysics 18, 1140-1145.

Ricciardi, L. M. and H. Umezawa, 1967, "Brain and physics of many-body problems, " Kybernetik 4, 44-48.

Richer, I. , 1965, "Pulse transmission along certain lumped non-linear transmission lines, " Electron. Lett. 1, 135-136.

Richer, I. , 1966, "The switch-line: a simple lumped transmission line that can support unattenuated propagation, " Trans. IEEE Circuit Theory, CT-13, 388-392.

Rinzel, J. , 1975a, "Spatial stability of traveling wave solutions of a nerve conduction equation, " Biophys. J. 15, 975-988.

Rinzel, J. , 1975b, "Neutrally stable traveling wave solutions of nerve conduction equations, " *J*. *Math*. *Biol*. *2*, 205-217.

Rinzel, J. , 1976, "Simple model equations for active nerve conduction and passive neuronal integration, " *Lectures* *Math*. *Life* *Sci*. *8*, 125-164.

Rinzel, J. and J. B. Keller, 1973, "Traveling wave solutions of a nerve conduction equation, " *Biophys*. *J*. *13*, 1313-1337.

Rinzel, J. , and W. Rall, 1974, "Transient response in a dendritic neuron model for current injected at one branch, " *Biophys*. *J*. *14*, 759-790.

Rissman, P. , 1977, "The leading edge approximation to the nerve axon equation, " *Bull*. *Math*. *Biol*. *39*, 43-58.

Roberts, G. E. and H. Kaufman, 1066, *Table* *of* *Laplace* *Transforms*, Saunders, Philadelphia, pp. 210 and 214.

Robertson, J. D. T. S. Bodenheimer, and D. E. Stage, 1963, "The ultrastructure of Mauthner cell synapses and nodes in goldfish brains, " *J*. *Cell*. *Biol*. *19*, 159-199.

Rochester, N. , J. H. Holland, L. H. Haibt, and W. L. Duda, 1956, "Tests on a cell assembly theory of the action of a brain using a large digital computer, " *Trans*. *IRE* *Inform*. *Theory*, *IT-2*, 80-93.

Rogers, C. R. , 1961, *On* *Becoming* *a* *Person*, Houghton Mifflin, Boston.

Rosen, G. , 1974, "Approximate solution to the generic initial value problem for nonlinear reaction-diffusion equations, " *SIAM* *J*. *Appl*. *Math*. *26*, 221-224.

Rosenberg, S. A. , 1969, "A computer evaluation of equations for predicting the potentials across biological membranes, " *Biophys*. *J*. *9*, 500-509.

Rosenblatt, F. , 1958, "The perceptron: A probabilistic model for information storage and organization in the brain, " *Psych*.

Rev. 65, 386-408.

Rosenblatt, F. , 1962, Principles of Neurodynamics, Spartan Books, New York.

Rosenbleuth, A. , 1941, "The stimulation of myelinated axons by nerve impulses in adjacent myelinated axons, " Am. J. Physiol. 132, 119-128.

Rosenbleuth, A. , N. Wiener, W. Pitts, and J. Garcia Ramos, 1948, "An account of the spike potential of axons, " J. Cell. Compar. Physiol. 32, 275-317.

Rosenfalk, P. , 1969, Intra- and Extracellular Potential Fields of Active Nerve and Muscle Fibres, Akademisk Forlag, Copen-hagen.

Roszak, T. , 1973, Where the Wasteland Ends, Anchor, New York.

Rozonoér, L. I. , 1969, "Random logical nets, " Avtomatika i Tele-mekhanika No. 5, 137-147; No. 6, 99-109; No. 7, 127-136 (translated in Automata).

Rushton, W. A. H. , 1951, "A theory of the effects of fiber size in medullated nerve, " J. Physiol. 115, 101-122.

Sabah, N. H. , and K. N. Leibovic, 1969, "Subthreshold oscilla-tory responses of the Hodgkin-Huxley cable model for the squid giant axon, " Biophys. J. 9, 1206-1222.

Samuel, A. L. , 1959, "Some studies in machine learning using the game of checkers, " IBM J. Res. Dev. 3, 210-223.

Sato, H. , I. Tasaki, E. Carbone, and M. Hallett, 1973. "Changes in axon birefringence associated with excitation: implications for the structure of the axon membrane, " J. Mechanochem. Cell. Motility 2, 209-217.

Sato, M. and T. Ishihara, 1974, "Graph theoretical approach to threshold systems, " Math. Jap. 19, 371-380.

Sato, R. , and H. Miyamoto, 1967, "Active transmission lines, " Electron. Commun. Jap. 50, 131-142.

Sattinger, D. H., 1976, "On the stability of waves of nonlinear parabolic systems," Adv. in Math. 22, 312-355.

Schadé, J. P. and J. Smith, 1970, Computers and Brains, Elsevier, Amsterdam.

Schmitt, F. O., P. Dev, and B. H. Smith, 1976, "Electrotonic processing of information by brain cells," Science, 193, 114-120.

Scott, A. C., 1962, "Analysis of nonlinear distributed systems," Trans. IRE CT-9, 192-195.

Scott, A. C., 1963, "Neuristor porpagation on a tunnel diode loaded transmission line," Proc. IEEE 51, 240.

Scott, A. C., 1964a, "Steady propagation on nonlinear transmission lines," Trans. IEEE Circuit Theory CT-11, 146-154.

Scott, A. C., 1964b, "Distributed device applications of the superconducting tunnel junction," Solid State Electron. 7, 137-146.

Scott, A. C., 1970, Active and Nonlinear Wave Propagation, Wiley-Interscience, New York.

Scott, A. C., 1971a, "Tunnel diode arrays for information processing and storage," Trans. IEEE on Systems, Man and Cybernetics, SM-1 267-275; see also U. S. Patent No. 3,822,381.

Scott, A. C., 1971b, "Effect of the series inductance of a nerve axon upon its conduction velocity," Math. Biosci. 11, 277-290.

Scott, A. C., 1972, "Transmission line equivalent for an unmyelinated nerve axon," Math. Biosci. 13, 47-54.

Scott, A. C., 1973a, "Information processing in dendritic trees," Math. Biosci. 18, 153-160.

Scott, A. C., 1973b, "Strength duration curves for threshold excitation of nerves," Math. Biosci. 18, 137-152.

Scott, A. C., 1976, "The application of Bäcklund transforms to physical problems," in Bäcklund Transformations (R. M. Miura, ed.) Springer, Berlin, Heidelberg, New York.

Scott, A. C., F. Y. F. Chu, D. W. McLaughlin, 1973, "The soliton: a new concept in applied science," Proc. IEEE 61, 1443-1483.

Scott, B. I. H., 1962, "Electricity in plants," Sci. Am. 207, No. 4, 107-117.

Scott Russell, J., 1844, "Report on waves," Proc. Roy. Soc. (Edinburgh), 319-320.

Shelley, M. W., 1818, Frankenstein (or a Modern Prometheus), Dell, New York (published in 1965).

Shimbel, A., and A. Rapoport, 1948, "A statistical approach to the theory of the central nervous system," Bull. Math. Biophys. 10, 41-55.

Singer, S. J. and G. L. Nicolson, 1972, "The fluid mosaic model of the structure of cell membranes," Science 175, 720-731.

Skinner, B. F., 1969, Contingencies of Reinforcement, Appleton-Century-Crofts, New York.

Skinner, B. F., 1971, Beyond Freedom and Dignity, Bantam/Vintage, New York.

Smith, D. R. and C. H. Davidson, 1962, "Maintained activity in neural nets," J. Assoc. Comp. Mach. 9, 268-279.

Smith, R. A., 1961, Semiconductors, Cambridge University Press, pp. 234-236.

Smolyaninov, V. V., 1969, "Speed of conduction of excitation along a fibre and a syncytium," Biophysics 14, 357-371.

Smolyaninov, V. V., 1970, "General theory of conduction of excitation in a medium, " Biophysics 15, 133-144.

Spencer, W. A. and E. R. Kandel, 1961, "Electrophysiology of hippocampal neurons: IV. Fast prepotentials, " J. Neurophysiol. 24, 272-285.

Steinbuch, K., 1961, "Die Lernmatrix, " Kybernetik 1, 36-45.

Strandberg, M. W. P., 1976, "Action potential in the giant axon of Loligo: a physical model, " J. Theoret. Biol. 58, 33-53.

Suzuki, R., 1967, "Mathematical analysis and application of iron wire neuron model, " Trans. IEEE Biomed. Eng. 14, 114-124.

Tasaki, I., 1939, "The electro-saltatory transmission of the nerve impulse and the effect of narcosis upon the nerve fiber, " Am. J. Physiol. 127, 211-227.

Tasaki, I., 1959, "Conduction of the nerve impulse, " Handbook of Physiology, Section 1, 75-121.

Tasaki, I., 1968, Nerve Excitation, A Macro-molecular Approach, Thomas, Springfield, Illinois.

Tasaki, I., 1974a, "Energy transduction in the nerve membrane and studies of excitation processes with extrinsic fluorescence probes, " Ann. N. Y. Acad. Sci., 227, 247-267.

Tasaki, I., 1974b, "Nerve excitation. New experimental evidence for the macro-molecular hypothesis, " in Actualités Neurophysiologiques, Masson, Paris, 79-90.

Tauc, L., 1962a, "Site of origin and propagation of spike in the giant neuron of Aplysia", J. Gen. Physiol. 45, 1077-1097.

Tauc, L., 1962b, "Identification of active membrane areas in the giant neuron of Aplysia, " J. Gen. Physiol. 45, 1099-1115.

Tauc, L. and G. M. Hughes, 1963, "Modes of initiation and propagation of spikes in the branching axons of molluscan central neurons, " J. Gen. Physiol. 46, 533-549.

Taylor, R. E., 1963, "Cable Theory," in W. L. Nastuk (ed.), Physical Techniques in Biological Research, Vol. 6, Ch. 4, Academic, New York, pp. 219-262.

Telesnin, V. R., 1969, "Stability of the steady state in a homogeneous excitable ring," Biophysics 14, 140-149.

Teorell, T., 1936, "Electrical changes in interfacial films," Nature 137, 994-995.

Thomas, R. C., 1972, "Electrogenic sodium pump in nerve and muscle cells," Physiol. Rev. 52, 563-594.

Tille, J., 1965, "A new interpretation of the dynamic changes of the potassium conductance in the squid giant axon," Biophys. J. 5, 163-171.

Trucco, E., 1952, "The smallest value of the axon density for which 'ignition' can occur in a random net," Bull. Math. Biophys. 14, 365-374.

Tsetlin, M. L., 1973, Automaton Theory and Modeling of Biological Systems, Academic, New York, 147-159.

Undrovinas, A. I., V. F. Pastushenko, and V. S. Markin, 1972, "Calculation of the shape and speed of nerve impulses," Doklady Biophysics 204, 47-50.

Vanderkooi, G., and D. E. Green, 1970, "Biological membrane structure, I. The protein crystal model for membranes," Proc. N. A. S. 66, 615-621.

Van Der Pol, B., 1926, "On relaxation oscillations," Phil. Mag. Ser. 7, 2, 978-992.

Van Der Pol. B., 1934, "The nonlinear theory of electric oscillations," Proc. IRE 22, 1051-1086.

Van Der Pol. B., 1957, "On a generalisation of the non-linear differential equations $u_{tt} - \epsilon(1-u^2)u_t + u = 0$," Proc. Acad. Sci. Amsterdam, A60, 477-480.

Van der Ziel, A. , 1957, Solid State Physical Electronics, Prentice-Hall, Englewood Cliffs, N. J. , pp. 445-449.

Van Heerden, P. J. , 1963, "Theory of optical information storage in solids, " Appl. Optics 2, 393-400.

Ventriglia, F. , 1974, "Kinetic approach to neural systems, " Bull. Math. Biol. 36, 535-544.

Von Euler, C. , J. D. Green, and G. Ricci, 1956, "The role of hippocampal dendrites in evoked responses and after discharges, " Acta Physiol. Scand. 42, 87-111.

Von Neumann, J. , 1958, The Computer and the Brain, Yale University Press, New Haven.

Von Senden, M. 1960, Space and Sight, Methuen & Co. , London. A translation of Raum-und Gestaltauffassung bei Operieten Blindgeborenen, published in 1932.

Vorontsov, Yu. I. , M. I. Kozhevonikova, and I. V. Polyakov, 1967, "Wave processes in active RC-lines, " Rad. Eng. Elec. Phys. , 11, 1449-1456.

Watanabe, A , and H. Grundfest, 1961, "Impulse propagation at the septal and commissural junctions of crayfish lateral giant axons, " J. Gen. Physiol. 45, 267-308.

Watanabe, A , S. Terakawa and M. Nagano, 1973, "Axoplasmic origin of the birefringence change associated with excitation of a crab nerve, " Proc. Jap. Acad. 49, 470-475.

Watson, G. N. , 1962, Theory of Bessel functions (2nd ed. ), Cambridge University Press.

Waxman, S. G. , 1972, "Regional differentiation of the axon: a review with special reference to the concept of the multiplex neuron, " Brain Res. 47, 269-288.

Waxman, S. G. and M. V. L. Bennett, 1972, "Relative conduction velocities of small myelinated and non-myelinated fibres in the central nervous system, " Nature (New Biology) 238, 217-219.

Waxman, S. G. and R. J. Melker, 1971, "Closely spaced nodes of Ranvier in the mammalian brain, " Brain Res. 32, 445-448.

Waxman, S. G. , G. D. Pappas, and M. V. L. Bennett, 1972, "Morphological correlates of functional differentiation of nodes of Ranvier along single fibers in the neurologenic electric organ of the knife fish Sternarchus, " J. Cell. Biol. 53, 210-224.

Weber, H. , 1873a, "Ueber die Besselschen Functionen und ihre Anwendung auf die Theorie der elektrischen Ströme, " J. f. d. reine und angew. Mathematik, 75, 75-105.

Weber, H. , 1873b, "Ueber die stationären Strömungen der Elektricität in Cylindern, " J. f. d. reine und angew. Mathematik, 76, 1-20.

Weinberg, A. M. , 1941, "Weber's theory of the kernleiter, " Bull. Math. Biophys. 3, 39-55.

Weinberg, A. M. , 1942, "Green's functions in biological potential problems, " Bull. Math. Biophys. 4, 107-115.

Weizenbaum, J. , 1972, "On the impact of computers on society," Science 176, 609-614.

Weizenbaum, J. , 1976, Computer Power and Human Reason, Freeman, San Francisco.

Weyl, H. , 1949, Philosophy of Mathematics and Natural Science, Princeton University Press, p. 282.

Wheeler, J. A. , 1974, "The universe as a home for man, " Am. Sci. 62, 683-691.

White, H. , 1961, "The formation of cell assemblies, " Bull. Math. Biophys. 23, 43-53.

White, S. H. , 1975, "Phase transitions in planar bilayer membranes, " Biophys. J. 15, 95-117.

Whitham, G. B. , 1974, Linear and Nonlinear Waves, Wiley-Interscience, New York.

Widrow, B. , and J. B. Angell, 1962, "Reliable, trainable networks for computing and control, " Aerospace Eng. , September, 78-123.

Wiener, N. , 1960, "Some moral and technical consequences of automation, " Science 131, 1355-1358.

Wilson, H. R. and J. D. Cowan, 1972, "Excitatory and inhibitory interactions in localized populations of model neurons, " Biophys. J. 12, 1-24.

Wilson, H. R. and J. D. Cowan, 1973, "A mathematical theory of the functional dynamics of cortical and thalamic nervous tissue, " Kybernetik 13, 55-80.

Yajima, S. , T. Ibaraki, and I, Kawano, 1968, "On autonomous logic nets of threshold elements, " Trans. IEEE Computers C-17, 385-391.

Yamaguti, M. , 1963, "The asymptotic behavior of the solution of a semi-linear partial differential equation related to an active pulse transmission line, " Proc. Jap. Acad. 39, 726-730.

Yoshizawa, S. , 1971, "Asymptotic behavior of a bistable transmission line, " IECE Professional Group on Nonlinear Problems, NLP 71-1, (in Japanese).

Yoshizawa, S. , and Y. Kitada, 1969, "Some properties of a simplified nerve equation, " Math. Biosci. 5, 385-390.

Yoshizawa, S. , and J. Nagumo, 1964, "A bistable distributed line," Proc. IEEE, 52, 308.

Young, J. Z. , 1936, "Structure of nerve fibers and synapses in some invertebrates, " Cold Spring Harbor Symp. Quart. Biol. 4, 1-6.

Young, J. Z. , 1951, Doubt and Certainty in Science, Clarendon Press, Oxford.

Yuan, H. T. and A. C. Scott, 1966, "Distributed superconductive oscillator and neuristor," Solid State Electron. 9, 1149-1150.

Zeeman, E. C., 1972, "Differential equations for the heartbeat and the nervous impulse," in C. H. Waddington (ed.), Toward a Theoretical Biology, 4,

Zeldovich, Y. B., and G. I. Barenblatt, 1959, "Theory of flame propagation," Combustion and Flame, 3, 61-74.

# Index